Int-I

INDICE PER ARGOMENTI -I

Questo INDICE è necessario dato che la materia trattata è scritta in modo irrituale dai testi classici Infatti da un lato presenta il fomulario delle leggi della fisica-matematica dei vari autori e le relative dimostrazioni nei limiti dello spazio ad esse riservato dalla pagina al seguito.Ma l'inserimento storico la riconduce alla narrativa :

Int-1

INDEX BY SUBJECT -I

CONTENTS This is necessary given that the subject matter is written in an amicable way from the classical texts fact, on the one hand presents the fomulario the laws of physics and mathematics of the various authors and their proofs within the limits of the space reserved to them-proved from the page to the following . But inserting the historic back to the narrative:

INDIZZAZIONE ↓

A

B

C

INDICE PER ARGOMENTI -II

INDEX BY SUBJECT-II

INDICE PER ARGOMENTI III — DEX BY SUBJECTS III

Dalla pg-46 alla pg-80
Indici [E]-[F]-[G]-[H]

A seguire → [H]

From pg -46 to 80
Index [E] - [F] - [G] - [H]

Sequency → H

INDICE PER ARGOMENTI IV

Dalla pg-81 alla pg-100

Indice [H]

INDEX BY TOPICS IV

From pg-81 to pg-100

Index [H]

DIMENSIONI DI CORPUSCOLI A CONFRONTO Fis3-2

Oggetto Denominazione	Dimensione metri	Dimensione assoluta	A confronto col DNA	linearizzazione indicativa
Protone	10^{-15}	1	0,000000000000001	1
Nucleo	10^{-14}	10	0,00000000000001	10
Atomo	10^{-10}	100.000	0,0000000001	100.000
Molecola Risonante	10^{-9}	1.000.000	0,000000001	1.000.000
DNA Molecola ...	10^{-7}	100.000.000	0,0000001	100.000.000

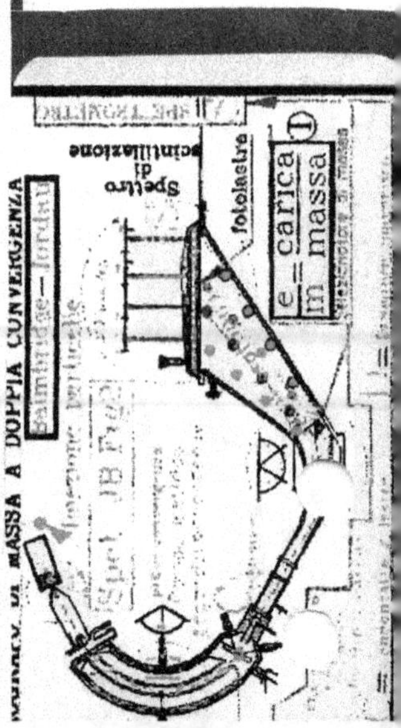

STORIOGRAFIA SCIENTIFICA A	STORIOGRAPHY SCIENTIFIC A
- Italiano-Inglese (Parte prima) -	**Italian-English (Part I) -**

I-Questo volume ha per oggetto la rievocazione storico-scientifica, da Babilonia al risveglio scientifico dell'occidente europeo. Con una personale suddivisione **Si parte dal popolo Caldeo** insediato fra il fiume Tigri ed il fiume Eufrate chiamato dai Greci Mesopotamia I Caldei erano esperti in astronomia come fa fede una tavoletta di terracotta rinvenuta nei pressi di Babilonia(**attuale Bagdad**) con effige della stella Sirio di magnitudine (luminosità) apparente di prima classe (attuale della metrica Pogson 0,2). La bibbia ricorda che il popolo ebraico dell'epoca era reso schiavo dei Caldei e che la stella Sirio aveva guidato i Re Magi verso il Redentore in Bethlemme . Il secndo periodo parte dall'anno zero dell'era Cristiana giungere al **XV secolo**. Comprende due eventi storici significativi scientificamente il 212 d.C.in corrispondenza alla caduta di **Siracusa** ad opera del proconsole **Marcello** . Questa data coincide con la morte di **Archimede**, si dice ad opera di un soldato romano che non riconoscendolo lo uccise .E' logico pensare che la condanna a morte sia stata pronunciata dal proconsole data la fama del personaggio e sopratutto perchè reo di aver inventato il primo cannone (Catapulta) .Con la quale lanciava sugli assalitori romani olio bollente.Il terzo dalla luce scientifica greca che si spegne, cioè dal 212 d.C al secolo XVdel RISVEGLIO SCIENTIFICO OCCIDENTALE

I-This volume has as its object the recalling historical and scientific, from Babylon to the scientific awakening of Western Europe.
With a staff division is part of the Chaldean people settled between the Tigris and the Euphrates river Greeks called Mesopotamia
The Chaldeans were skilled in astronomy as an authentic terracotta tablet discovered near Babylon (**now Baghdad**) with the effigy of the star Sirius magnitude (brightness) apparent first-class (current metric Pogson 0.2). The Bible points out that the Jewish people at the time was made a slave of the Chaldeans, and the star Sirius had guided the Magi to the Redeemer in Bethlehem. The secndo period starts from the zero of the Christian era to reach the **fifteenth century**. Includes two significant historical events scientifically the 212 d.C in correspondence to the fall of **Syracuse** by the proconsul **Marcello**. This date coincides with the death of **Archimedes**, it is said by a Roman soldier who is not intent-recognisin killed him. E 'logical to think that to sentence to death has been pronounced by the proconsul given the fame of the character and especially because the offender to have invented the first cannon (Catapulta.) with which threw oil on the attackers Roman bollente.
Il third from the Greek scientific light that goes out, that is, from 212 C.d to the fifteenth century of VAKING UP WESTERN SCIENTIFIC

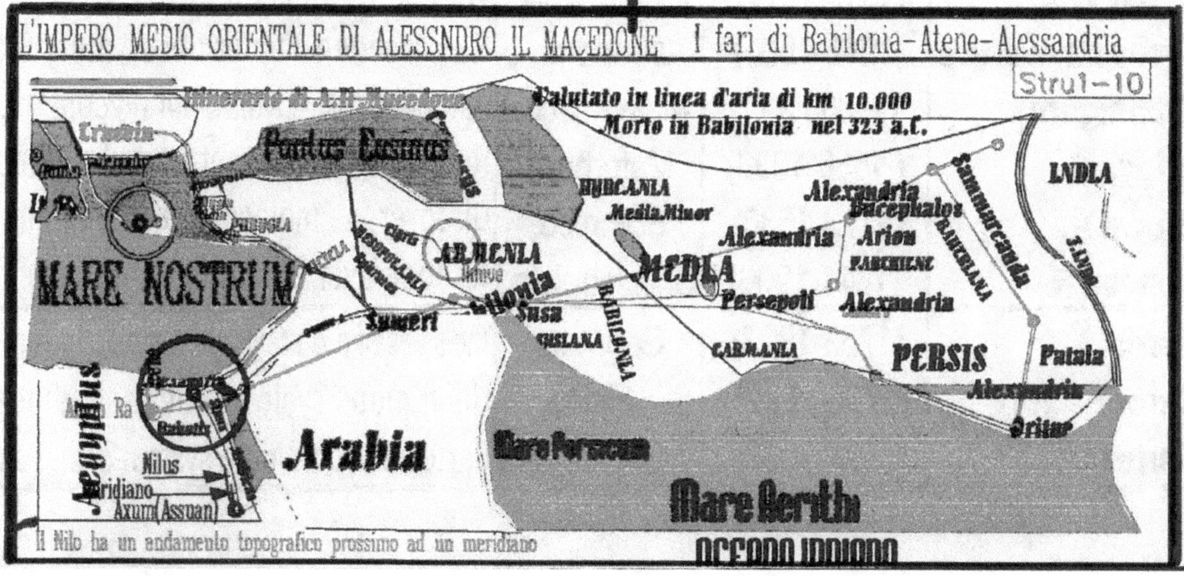

AUTORE	VISSUTO	Operatore	Main works of the author
Abu'Hasan Tabit	~950 d.C.	Geometra	Algebrista and surveyor
Abu'Kamil.....	~950 d.C.	Algebrista	Received by means of the Arabic Al Mgesto
Abu'(Albategno)	~(858-929) d.C.	Astronomo	Trigonometrista and the cosine law
Aryabhata	~(598 d.C.?)	Matematico	Author of the Indian text "Sulvasutras"
Appolonio di Pergia	~(170.a.C)	Geometra	Author of the method of conics from cone
J.M.Ampere	(1775-1836)	Elettrofisico	The idea of the flow of electrons as current
Archimede di Siracusa	(287-212 a.C)	Fisico-matemtico	The squaring of circle, mathematical
Archita di Taranto	~500 a.C.	Astronomo	Astrophysics. He ruled with good laws
Aristarco di Samo	~3o sec.a.C	Astrononmo	The ancient Copernicus and his Shpera fixaru
Aristotile(Stagira)	(384-322 a.C.)	Filosofo	Founder of the Lyceum opposed Academy
Aston W.A.	(1877-1945)	Ricercatore	He built the mass spectrometer for isotope
Avogadro A. conte	(1776-1856)	Ricercatore	$A=6,2510^{23}$=molecules (in one liter to 293°K
Bacone R.	(1214-1294)	Filosofo	Doctor Mirabilis (I think therefore I am)
Bequerel H.	(1852-1908)	Fisico	He discovered the radioactivity of uranium (U^{23}
Bernouilli D.	(1700-1782)	Fisico	He theorized the motion of real fluids
Bessel di Menden	(1784-1846)	Astrofisico	He solved the problem of stellar distances
Biot J.B.	(1774-1862)	Fisico	Among the various law of electro-magnetis
Bohr N.	(1885-1962)	Fisico	Its the law of emission of photons
Boltzmann L.	(1844-1906)	Fisico	Introduced the concept of entropy in the TD
Boaga G.	(1902-1961)	Geodeta	Introduced the Gauss-Boaga coordinated
Boyle R	(1626-1691)	Chimico	Isotherms of a real gas in TD
Black J.	(!728-1799)	Fisico-chimico	Theorized specific heats of gases
Bradley J.	(1693-1726)	Astronomo	Calculated the stellar parallaxes to Greenwic
Bragg H.W	(1862-1942)	Costruttore	He was the first to realize a spectrometer
Bragg W.L	(1890-1971)	figlio di **H.W**	Worked with his father for the realization
Brahe di Knudstorp	(1546-1601)	Astronomo	He preceded Kepler observatory Prage
Brahmgupta	(~800 a.C)	Matematico	Author of the "Livalati" of algebra and geomet
Briggs E.	(1556-1630)	Matematico	Relations trigonometric and exponential
Brown R.	(1773-1630)	Botanico	Discover the molecular motion in the emulsio
Bunsen R.	(1811-1906)	Chimico-fisico	He took over the spectra of flame in a crue
Carnot L.	(1753-1823)	Geometra	The sides of a Δ are proportional to the opposite cosi
Carnot Sadi	(1796-1832)	Fisico	His famous cycle-isothermal adiabatic
Cartesio R	(1596-1630)	Geometra	Cartesian coordinate system orthogonal (C.C.

AUTORE	VISSUTO	Operatore	Opere principali dell'autore
Cartesio R.	(1596-1650)	Geometra	Il sistema di coodinate ortogonali(c.c.o)
Cassini G.D	(1625-1712	Astrononomo	Determinò le effemeridi delle stelle
Cauchy A.L	(1789-1857)	Matematico	Teorico delle funzioni reali e complesse
Capteyn J.	(1854-1922)	Astrofisico	La nostra Gallassia è un ellissoide di rotazione
Chadwich	(1891-1974)	Fisico	Separò α e β emessi dal nucleo di (U^{238})
Clausius R.E.	(1822-1888)	"	Suo il 2^0 principiodella termodinamica
Copernico N	(1473-1543)	Astronomo	Non è la Terra ferma ma il Sole(Sol Stat)
Compton A	(1892-1962)	Fisico	Scoprì l'emissione elettroni fotoni (hf)
Coulomb C.A	(1736-1806)	"	Sua la legge su le interazioni fra cariche
Crooker W.	(1879-1918)	"	Sccprì i raggi X e costruì il tubo catodico
Curie P.	(1859-1906)	"	Il Radio ,elemento del nucleo,dall'Uranio
Dalton J.	(1766-1844)	Chimico	Enunciò la legge delle proporzioni
De Sitter W.	(1872-1932)	Astrofisico	Calcolò il raggio fisico dell'universo
De Broglie L.	(1892-1973)	Fisico	Fondatore della meccanica ondulatoria
De Moivre	(1667-1744)	Matematico	Formulò lalegge dei numeri comlessi
Dirac P.A.	(1902-1984)	Fisico	L'elettrone ed il suo spin magetico
Einstein A.	(1879-1955)	Fisico	Dal quadrivettore alla energia=massa
Eratostene	(276-195 a.C.)	Geometra	Determinò per primo il raggio R della Terra
Euclide(di Megara)	(~ 300 a.C.)	"	Legislatore dello spazio lineare=euclideo
Eudosso(diCnido)	(~ 450 a.C.)	Astronomo	Ideò l'universo come Sphera fixarum
Eraclito(di Efeso)	(540-480 a.C)	Filosofo	L'elemento primordiale è il fuoco
Faraday M.	(1791-1867)	Fisico	Scpore l'induzione magnetica e lo ione
Fermat P.	(1601-1665)	"	Sua la legge di riflessione ottica
Fizeau I.L.	(1819-1896)	"	La misura della velocitàdella luce c
Foucoult L.	(1819-1868)	"	Scoprì le correnti parassite indotte
Fourier G.B	(1768-1830)	Matematico	Le serie impulsive di fotoni dell'elettrone
Fresnel A.G.	(1788-1827)	Fisico	Consolida la teoria ondulatoria della luce
Fraunhofer J	(1787-1826)	Ricercatore	Predisse la composizione del Sole
Galileo Galiei	(1564-1642)	Fisico	Gravità ,isocronismo,satelliti di Giove
Gamow G.A.	(1904-1968)	Astrofisico	La molecola di H origine dell'universo
Gauss C.F.	(1778-1855)	Matematico	I numeri complesi z=x+jy e geodetica s
Gay Lussac	(1778-1850)	Fisico	La isocora (V=Cost) dei gas perfetti
Geiger H.W.	(1882-1945)	Ricercatore	Ha realizzato il contatore di particelle

AUTORE	VISSUTO	Operatore	Main works of the author
Hamilton W.	(1805-1865)	Astronomo	League classical mechanics to quantum
Hansen P.A.	(1795-1874)	Geodeta	Inaccessible side of the quadrilateral solved
Helmholtz H	(1821-1894)	Fisico	Maxwell's equations and the potential
Hertz H.	(1857-1894)	"	Wave propagation elettomagnetiche
Hubble E.	(1889-1953)	Astrofisico	Discovered the dynamics of the physical universe
Huygens C.	(1619-1695)	Fisico	Light is a wave phenomenon
Ibn Haitane al....	(965-1039)	Astronomo	The earth does not move with uniform
Inn Sina......	(980-1037)	Matematico	Geometry and the law of perfection
Ipparco di Nicea	(~ 280 a.C.)	Astronomo	First, measured the Earth-Moon distance
Jeans J.H.	(1877-1946)	Fisico	The correct stellar radiation by Planck
Joule J.P.	(1818-1889)	"	Established the equivalence of work and heat Q A
Joung J	(1799-1878)	"	The magnetic induction particularized
Keplero W.J.	(1571-1630)	Astronomo	Legislator of the solar planetary system
Kircchoff G.R.	(1824-1887)	Fisico	The current-voltage electricity network
Kronecker L.	(1823-1891)	"	Equations and the solution of systems
Lagrange G	(1763-1827)	Matematico	Cordinated of specific details laws of fluid
Laplace P.S.	(1749-1827)	Astrofisico	Dall'ellissoide land to heavenly
Lavoiser A.L.	(1743-1794)	Chimico	Born sweeping chemistry alchemy
Legendre A.M.	(1752-1833)	Astrofisico	The spherical triangle heavenly fundamental
Lenz E.C.	(1804-1912)	Fisico	Laws of the field induced current functions
Laurent P.H.	(1841-1908)	Matematico	Complex variable
Lochyar S.N.	(1836-1920)	Fisico	Discoverer (2eHe4 helium) isotopic
Lorentz H.A.	(1853-1928)	"	Law electron motion with respect to the core
Jeans J.H.	(1877-1946)	"	Spectrum radiant high frequencies
Joung J.	(1799-1878)	Ricercatore	Experimented with the first induction motor b
Joule J.P.	(1818-1889)	Fisico	He established the equivalence heat Q and work A
Kircchoff H.A.	(1824-1887)	"	Principles of the current node and tensions
Kapteyn J.C.	(1851-1922)	Astrofisico	Our galaxy is a rotating ellipsoid
Kroneker L.	(1823-1891)	Matematico	Algebraic structures and elliptic functions
Meyer R.	(1814-1875)	Fisico-chimico	Principles 1st and 2nd law of thermodynamics
Malus E.L.	(1775-1812)	Fisico	Birefringent light wave is bipolar
Maxwell J.K.	(1831-18[illegible])	"	Electromagnetic space of campi
Michelson A.	(1852-1931)	Ricercatore	With the interferometer saw the Terra Firma

N	AUTORE	VISSUTO	Operatore	Main works of the author
100	Max Von Laue	(1879-1960)	Fisico	Discovered the diffraction optical prismatic
101	Millikan R.A.	(1868-1953)	"	Measured the electron charge and gravity
102	Monge G.	(1746-1818)	Matematico	Projective space and the objective images
102	Morley E.W.	(1838-1923)	Ricercatore	Helped Michelson in the famous experience
103	Moseley H.G.	(1887-1915)	"	Emissions A and B of uranium (U238)
104	Nagaoka H.	(1865-1965)	Fisico	Electronic spin=the nucleousSpirit Natural
105	Nepero J.	(1550-1617)	Matematico	Natural Logarithms of base e ($\ln x = e^x$)
106	Newton I.	(1643-1727)	Fisico	Universal attraction of matter (f=ma)
107	Hertz H.	(1857-1894)	Ricercatore	The test of the fields k, h electromagnetic
108	Hertz G.	(1887-1975)	"	Nobel Prize in 1955. The gas diffusion
109	Heaviside O.	(1850-1925)	Fisico	Layer impedance Z Earth
110	Pacinotti A.	(1841-1912)	Fisico	Prototype (1859) of the dynamo or motor
111	Parceval M.A.	(1755-1836)	Matematico	Lines harmonics of Fourier integral
112	Pauli W (Von)	(1900-1958)	Fisico	Reaction e^+ +e-> y (exclusion principle)
113	Papin D.	(1647-1714)	"	The thermodynamics of the steam valve
114	Picard E	(1620-1682)	Matematico	Figured the geodesic Amines-Malvoisine
115	Pitagora (di Samo)	(586-500 a.C)	"	Founded his own school. Known for theorem
116	Plank Max	(1858-1947)	Fisico	Born quantum physics of radiation
117	Platone (di Megara)	(427-347 a.C)	Filosofo	Founder of ethics and idealism
118	Poincarè H.	(1854-1912)	Matematico	Suggested to Lorentz the existence of the spin
119	Poisson S.D.	(1871-1940)	Fisico	Theory of heat and many potential fields
120	Rayleigh J.W.	(1842-1919)	"	Search quantum (Nobel 1904). Find the light
121	Regiomontano	(1436-1476)	Astronomo	(Muller) Resolution of the spherical triangle
122	Ritz W.	(1898-1947)	Ricercatore	Expanded Bohr spectrum x thermogenesis
123	Rö emer O.	(1644-1710)	Fisico	From the Earth-Sun constant distance
124	Rö ntgen W.H.	(1845 -1923)	Ricercatore	X-rays of the core aep (U^{238})
125	Rutherford E.	(1871-1937)	Fisico	The electron describes an elliptical orbit
126	Schwarschild K.	(1873-1916)	Astrofisico	Theory photometric (stellar atmospheres)
127	Sclodowska ?	(1887-1932)	Fisico	Discovered the Torio radioactivity ($90Th^{232}$)
128	Schrödinger E.	(1887-1951)	"	The function wave of the electron
129	Socrate di ?	(469-399 a.C)	Filosofo	The cult of right. Therefore assassinated!
130	Soddy F.	(1887-1956)	Fisico	Investigated the nuclear reactions of isotopes
131	Sommerfel A	(1868-1951)	"	He called the azimuthal quantum electron

N	AUTORE	VISSUTO	Operatore	Main works of the author
132	**Snellius W.**	(1581-1626)	Geodeta	The trigonometric series as geodetic
133	**Stark J.**	(1874-1957)	Fisico	The law of refraction ca electromagnets
134	**Stephenson ?**	(1781-1844)	Ricercatore	The thermodynamics and the first locomotive
135	**Stokes G.G.**	(1819-1903)	Fisico	Wrote on fluorescenza, dif and birifrazioni
136	**Tacquet A.**	(1612-1660)	Matematico	He developed, first, the geometric series
137	**Talete (di Mileto)**	(639-548 a.C)	"	Based on Euclidean similarity
138	**Thomson W.**	(1824-1907)	Fisico	Allias Kelvin and the absolute temperature °K
139	**Thomson J. J.**	(1856-1940)	Ricercatore	Translated Almagestadal from Arabic edition
140	**Tolomeo C.**	(~90-168d.C)	Astronomo	Optics in 5 books-The Earth center of cosmos
141	**Van der Waals**	(1837-1923)	Ricercatore	Formulated the theory of an ideal gas pV=RT
142	**Viete F.**	(1540-1603)	"	Inventor of the algebraic properties logicae
143	**Volta A.**	(1745-1827)	Fisico	Inventor of the electric battery or the emf
144	**Watt J.**	(1736-1819)	"	Made operational pV = RTcon the valve ap
145	**Weber W.**	(1804-1890)	"	Current as a flow of electrons
146	**Werner G.**	(1468-1527)	Matematico	Trigonometry and relation sin-cos-in
147	**Wien W.**	(1864-1928)	"	Energy E (X) has no spectrum X> 480n.m
148	**Wilson T.**	(1869-1959)	Fisico	The room to see the invisible
149	**Zeeman P.**	(1865-1943)	"	Spectrum of particles in the range
			Fuori indice	
150	**Mercatore G.**	(1512-1584)	Cartografo	He built the charts and the rhumb
151	**Heaviside O.**	(1850-1916)	Fisico	Founded the symbolic computation
152	**Rodrigues O.B.**	(1794-1851)	Matematico	Polinomio of Legendre normato P(z) = 1
153	**Ohm.G.S.**	(1787-1854)	Fisico	Law of electric currents: i = E/R
154	**d'Alembert**	(1717-1783)	(Jean le Rond)	Physicist. Law of the reaction of inertia f=am
155	**Hertzsprung**	(1783-1967)	Astronomo	With Russell discovered the nature of stellar gas

Hertzsprung-Russel -> Diagram Gotrìan: Fig40

If you have been told that copy is a trivial matter they have been deceiving. All the great effort that we ha made the list (sorry for the forgotten, to whom we have dedicated one (extra space index but only part the deserving but has found a place.) On the other hand copying can become fraught with difficulties. This proved by the works listed in the above-mentioned authors. Forced to copy what his predecessors had writt (the time that passes is an implacable judge consolidates or demolishes.) To then conclude with their person vision of the problems already treated by others. We instead had to laboriously copy for another reason indispensable prologue of scientific historiography. Without this guide we will be incurred and when in dou of who the credit? That the partial indexes for authors provides a reliable reference.

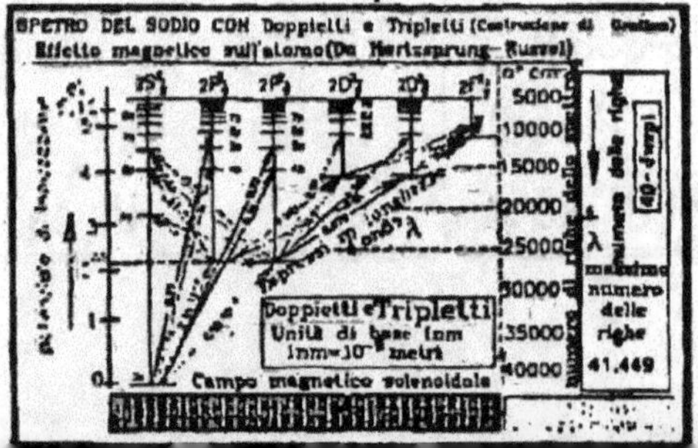

Int-I

INDICE PER ARGOMENTI -I

Questo INDICE è necessario dato che la materia trattata è scritta in modo irrituale dai testi classici Infatti da un lato presenta il fomulario delle leggi della fisica-matematica dei vari autori e le relative dimostrazioni nei limiti dello spazio ad esse riservato dalla pagina al seguito.Ma l'inserimento storico la riconduce alla narrativa :

Int-1

INDEX BY SUBJECT -I

CONTENTS This is necessary given that the subject matter is written in an amicable way from the classical texts fact, on the one hand presents the fomulario the laws of physics and mathematics of the various authors and their proofs within the limits of the space reserved to them-proved from the page to the following . But inserting the historic back to the narrative:

INDICE PER ARGOMENTI -II

INDEX BY SUBJECT-II

La Cina nota come il celeste Impero è famosa per aver scoperto la bussola, i carri magnetici e i caratteri di stampa mobili.Da questo è lecito dedurre che era una moltitudine di popoli predisposta a considerare il lato pratico delle cose I carri con bussola servivano ai loro imperatori per riscuotere le imposte nello sterminato impero.

Nel XII secolo **Marco Polo** seguito poi dai padri gesu- iti in quel lontano paese ne dice un gran bene per la accoglienza ricevuta dagli abitanti delle regioni visitate. Sappiamo che i Cinesi hanno dato un inirizzo praticistico alle loro nozioni di aritmetica . I numeri venivano sommati usando dei bastoncini di bambu . Fra le discipline scientifiche la geometria èstata certamente la più coltivata a fini principalmente pratici .Ad esempio come disegnare una piazza avente per limite i lati di un quadrato.Di certo conoscevano che i lati di un quadrato soddisfacevano alla relazione $3^2+4^2=5^2$ (1) .

Non giusero alla generalizzare quello che **Pitagora** formulò per il rettangolo fra lati e diagonale secondo

$$d^2=a^2=b^2+c^2 \qquad (2).$$

Il cinese Sun Yang autore di una aritmetica rappresenta i numeri 1.2,3, come detto con asticelle di bambu simili ai numeri romani I-II-III

La geometria di **Yang** è stata scritta nel 550 (a.C.). Riporta la soluzione del problema << Un gallo costa 5 monete ed una gallina 3 ,mentre tre polli valgono una moneta, Con 100 monete si sono aquistati esattamente 100 bipedi .La domanda è ,quanti, di ciscuna specie?Un algebrista scriverà le condizioni -:

- $x+y+z=100$ (3) $5x+3y+\frac{z}{3}=100$ (4) -

Dato che le incognite x,y,z sono 3 per risolvere il problema occorre una terza condizione>>

Diversamente si dovrà assegnare ad x,y,z dei valori per tentativi Questo ha fatto Yang che dà i valori :

- $(x=4, y=18, z=28)$

Chan Heng quadrò il cerchio ottenendo la miura $\pi=3,16227$.con 2 succeioni convergeni di lati di un poligono inscritto e circoscritto. Equazioni fino al quarto grado sono proposte . Non risolte in forma generalizzata. Per questo dovremo aspettare l'Italiano **N.Fontana**(Tartaglia) plagiato e truffato da **Cardano**

China known as the Celestial Empire is famous for his discovery of the compass,magnetic car and fonts mobili.Da this can be inferred that it was a multitude of nations prepared to consider the practical side of things I needed to their chariots with compass emperors to collect taxes in empire. In the twelfth century **Marco Polo** followed by fathers jesus-ites in that distant country it says a lot of good for the welcome received by the inhabitants of the regions visited. We know that the Chinese have given a praticistico to their notions of arithmetic. The numbers were added together using bamboo sticks. Between the sciences geometry Estata certainly the most cultivated mainly practical purposes. Example how to draw a square having sides of a limit for quadrato.Di certainly knew that the sides of a square satisfy the relation $3^2+4^2=5^2$ (1)

Not giusero to generalize what **Pythagoras** formulated for the rectangle between the sides and diagonal second $d^2 = a^2 + b^2 = c^2$ (2). The Chinese Sun Yang is the author of an arithmetic numbers 1.2,3, as mentioned with strips of bamboo courtly Roman I-II-III. The geometry of Yang was written in 550 (BC). Returns the solution of the problem << A rooster and a hen costs 5 coins 3, while three chickens are worth a coin, with 100 coins were exactly 100 aquistati bipeds. The question is, how many, to draw each species? Algebraist A write conditions $x+y+z=100$ (3) $5x+3y+\frac{z}{3}=100$ (4) - Since the unknowns x, y, z are 3 to solve the problem, a third condition >> Otherwise you will have to assign to x,y, z valu- es for attempts This did Yang which gives the values: $(x = 4, y = 18, z = 28)$. **Chan Heng** quadrò the circle getting miura $\pi = 3.16227$. with successios with two of the sides of a polygon inscribed and circumscribed. Equations up to the fourth degree are proposed. Unresolved in generalized . For this we will have to wait for the Italian **N.Fontana**(Tartaglia) plagiarized and cheated by **Cardano**

LA GEOMETRIA DEI CINESI

PITAGORA 570-495 a.c.

Il S.P.QR domina incontrastato per decidre dei destini del mondo allora conosciuto.
Il Suo dominio si estendeva dai mari del Nord Europa ai deserti della Mesopotamia . Ma la potenza temporale segue una legge inflessibile. Dove c'è potenza c'è ricchezza e con questa inseparabile e servizievole ancella genera corruzione.
Gli storici ancora una volta non son credibili. Affermano che Roma crollò a cauda delle invasioni barbariche. Al contrario i "barbari delle tribù indogermaniche" erano un serbatooio al servizio del S.P.R. pronti a difendere i confini dello steminato impero. Non solo il Colosseo (circenses) ma pure il pane . Frutto spesso della vendita al miglior offerente delle Prefetture sui tre continenti. Vendute al miglior offerente su base delle potenziali concussioni realizzabiuli , rapporto ricchezza territorio da soggiogare. La spartizione dei proventi illeciti realizzati dal degrado dei discendenti di Cesare " Alea Jacta" est" Un solo Dio, il denaro" che per diritto divino appartienee ai potenti ed ai ricchi. Cosa sta faccendo quell'Uomo in quel di Israel? Sta predicando la eguaglinza tra ricchi e poveri.! SPQR ordina al Sinedrio "uccidetelo" e se continua crociffigetelo! Che serva da esempio ai pover che vogliono il denaro rubando ai ricchi.Ma il morbo si diffondeva proprio per la crocefissione. La falange dei martiri si ingrossava fino aprovocare il crollo di Roma.
Il primo Apostolo Pietro trovò la consacrazione in Caput Mundi ora Roma Urbi et Orbi Ma il destino sta nel D.N.A. dei popoli e è inesorabile.
Si nasce innocenti e senza peccato ed è più facile non morire tali. Ciò non vale per il potere perchè fatalmente arriva la catastrofe .Una mela marcia contamina le sane. Sotto questa prospettiva vediamo i mala tempora. La inquisizione vigila ed incrimina e il Santo Ufficio, l'organo giudicante , condanna Ne fanno le spese gli scienziati Copernico in primis, indiciato per Sol Stat di cui diremo

The SPQR dominates decidre for the destiny of th then known world. His dominion extended from th seas of Northern Europe to the deserts of Mesopotamia. But the temporal power follows an inflexible law. Where there is no power and wealtl with this non-detach-and obliging servant generate corruption. Historians once again I am not credible They say that Rome fell to the cauda of the barbari inva-sions. On the contrary, the "barbaric tribes indogermaniche" were a serbatooio serving the SP ready to defend the borders of the empire steminat Not only is the Colosseum (circuses) but also the bread. Often the result of the sale to the highest bidc of the three prefectures continenti. Vendute to the highest bidder on the basis of potential concussion realizzabiuli, ratio of wealth to the territory to be subdued. The division of criminal proceeds realized the degradation of the descendants of Caesar "Ale Jacta" east "One God, the money" that by divine ri appartienee the powerful and the rich. What is tha Man making as in that of Israel? Is preaching the eguaglinza between rich and poor.! SPQR Sanhedr order to "kill him" and if it continues crociffigetel To serve as an example to pover who want the mon stealing from the ricchi.Ma the disease was spreadi because of the crucifixion. The phalanx of the marty swelled up aprovocare the collapse of Rome. The fi Apostle Peter found the consecration in Rome Cap Mundi now Urbi et Orbi But fate is in the DNA o peoples and is inexorable. We are born innocent an without sin and it is easier not to die such. This is n true of the power because it inevitably comes the catastrophe. A rotten apple contaminates healthy. Under this perspective we see the mala tempora. Tl inquisition monitors and prosecutes and the Holy Office, the hearing body, condemns We are payin scientists , Copernicus first, indiciato for Sol St which we will discuss

Questo popolo dislocato ai confini dell'India non aveva fatto parlare di sè stesso .Di colpo si desta dal lettargo , nello spirito e nel cprpo, alla voce del profeta Maometto (571-643(d.C.)) .Invade le Indie e connquista conquista la Persia con Babi-lonia (Bagdad) e dilaga nella Mesopotamia.

Con una marcia inarrestabile ,simile a quella di A. Magno il Macedone. Attraversa il Sinai e co-me un'onda di marea dilaga in tutti i paesi della sponda mediterranea a comniciare dall'Egitto e cosi via sino allo stretto di Gibilterra.

Si trasferiscepoi sulla sponda orientale del Mare Nostrum soggiogando il portogallo . Non sazio di tanto dominio conquista la Spagna meridiona-le, la Sicilia . Prosegue la marcia invadendo la Grecia risalendo l'Illiria . Infine forse esauto per la lunga cavalcata viene sconfitto dal generale tedesco **Scandemberg**. Non esiste un esempio storico che un impero di quelle proporzioni si disfacesse in meno di due secoli. Ci interessa la cultura degli Arabi più che le scimitarre ispirate da Maometto del colosso dai piedi di argilla.

La loro opera in algebra (Al Gbr)e geometria ci ricordano che, con la protezione del califfo **Al Mamum** (813-833)venne eseguita una misura di un arco di meridiano, dopo quella più famosa di **Eratostene**(276—98 a,.C.). Insomma intorno al X secolo gli arabi raccolsero quanto di cultura trovarono a Bagdad e ad Alessandria di Egitto (La più famosa biblioteca del mondo)diretta allo-ra da un Arabo. Sono state importate dagli Arabi le funzioni trigonometriche seno(Q-P) e coseno (O-Q)di matrice indiana erano note fin da allora

Fi6-Am1,in versione moderna .

LA ARITMETICA DEGLI ARABI

Per scrivere gli interi da 1 a 400 gli Arabi ,come i Greci si servivano delle lettere a,b,...,z ; A,B,...,Z e più semplicemente con le dita delle mani . Ma **Allah** è grande.Ha donato loro lo 0 (gobar=ciffra)e a noi il sistema decimale .

This people stationed on the borders of India had not done talking about himself. Suddenly she awakened from lettargo, in spirit and cprpo, the voice of the Prophet Muhammad (571-643 (AD)). Invade the Indies and the conquest of Persia connquista with Babi-lonia (Baghdad) and ramps in Mesopotamia. With a relentless march, similar to that of A. The Great of Macedon. Through the Sinai and co-me a tidal wave is rampant in all the countries of the Mediterranean comniciare Egypt and so on up to the Straits of Gibraltar. It trasferiscepoi on the eastern shore of the Mediterranean subjugating portugal. Not satisfied from time domain conquers the States Southern Spain, Sicily. Continue running invading Greece going Illyria. Finally, perhaps long exhausting for the long ride is defeated by the German general **Scandemberg**. There is no historical example of an empire that those proportions fell apart in less than two centuries. We are interested in the culture of the Arabs more than the scimitars inspired by Muhammad's colossus with feet of clay. Their work in algebra (Al Gbr) geometry and remind us that, with the protection of Caliph **Al Mamum** (813-833) was performed a measurement of an arc of the meridian, after the most famous of **Eratosthenes** (276Å98 a,. C.). In short, around the tenth century the Arabs picked up the culture found in Baghdad and Alexandria in Egypt (The most famous library in the world) to direct-ing from an Arab. Were imported by the Arabs the trigonometric functions sine (Q-P) and cosine (O-Q) of Indian background were known since then **fi6-Am1, in a modern version.**

THE ARITHMETIC OF ARAB

To write the integers from 1 to 400 the Arabs, as the Greeks made use of the letters a, b, ..., z, A, B, ..., Z, and more with just your fingers. But **Allah** is grande.Ha given them the 0 (gobar = ciffra) and we are the decimal system

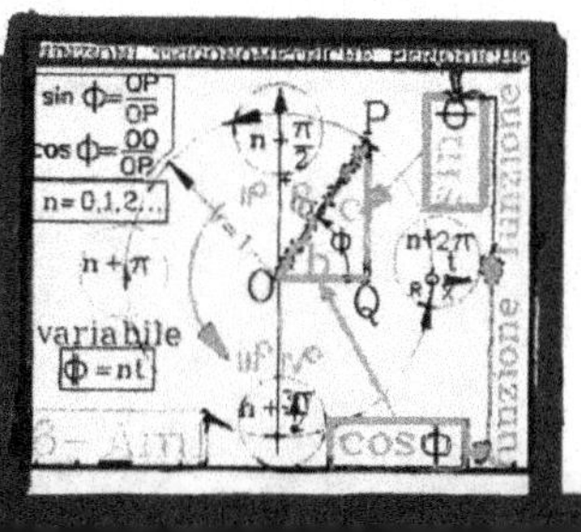

I popoli orientali e medio orientali usavano le lettere alfabetiche per indicare i numeri naurali interi da 1 a 400. Gli Arabi ,gli Ebrei e gli stessi Greci. Il matematico , astronomo , **Pitagora,** al secolo Πψτηαγορασ di Samo (571-570 a,C,) Metaponto(497-496) . fondatore di una scuola ha formalizzato il Teorema : $a^2 = b^2 + c^2$ (1) con Riferimento geometrico di **Fisc2-Fig10** .

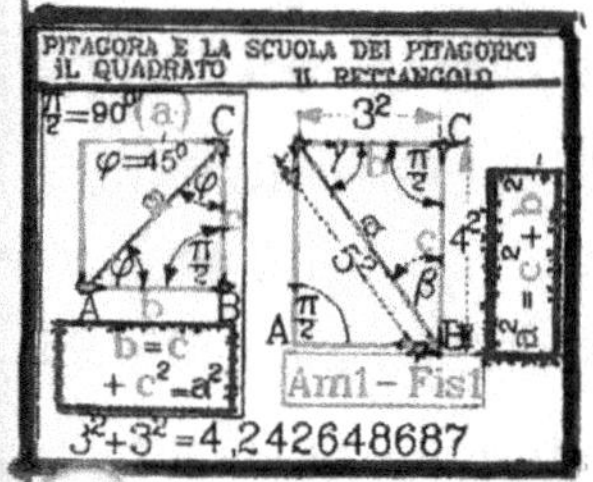

Si osservi che ,nicchia (a) , che il triangolo (definito dai tre lati a, b,c e tre angoli opposti α,β γ) ha , per costruzione i tre angoli : $\alpha=90^o \equiv \pi/2$; $\beta=\gamma=45^o \equiv \pi/4 \in$ (a)

Il problema della risoluzione

<<Dato un triangolo piano si dice che è risolto se, dati tre elementi di cui uno almeno un lato, si trovano gli altri tre elementi>>

Se ne deve dedurre che dati gli elementi ∈ (a) , ossia i soli tre angoli, il trian golo non è risolvibile dato che esistono infiniti triagoli simili che hanno gli stessi angoli ma lati diversi. Qusto è appunto il teorema caso di **Pitagora Infatti la soluzione consiste nel miurare due. lati** b e c e se b = c si ha : $2b^2 = a^2$ (1) da cui risulta : $a = b\sqrt{2}$ (2) . Resta da precisare che i geometri (**Pitagora**) e i geodeti(**Eratostene**) dell'epoca, usavano la stadia (circa 2,8 metri) per misurare le dis-tanze ed il sestante , **Fig 27-Strum**, per misurare gli angoli.

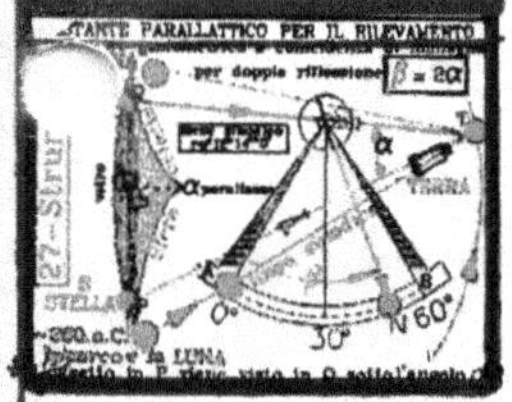

Usato da **Ipparco** per misurare la distanza Terra- Luna.Nel caso (1) la scuola pitagorea , di Metaponto , era in grado di tracciare sul terreno dei trinaoli rettangoli e quindi di costruire i triangoli rettangoli equilateri (a) e i triangoli rettangoli qualunque (b). La differenza ai fini della risoluzione è fondamentale. Infatti (a) si risolve con l'aritmetica (2) mentre (b) richiede gli operatori trigonometrici. Infatti si supponga di conoscere a=120 m, $\beta=38^o28'$. Allora per il Teorema di **Pitagora** possiamo scrivere ....

The Eastern peoples and Middle Eastern countries used the letters of the alphabet to denote numbers containing natural integers from 1 to 400. The Arabs the Jews and the Greeks themselves. The mathemati cian, astronomer, Pythagoras, born Pythagoras of Samos (571-570 a, C) Metaponto (497-496). Founder of a school has formalized Theorem: $a^2 + b^2 = c^2$ (1 with reference geometric **Fisc2- fig10.** Note that, niche (a), that the triangle (defined by the three sides a, b, c and three angles opposite α,β γ) to has, for building the three angles: $\alpha = 90^o = \pi / 2$, $\beta = \gamma = 45^o \equiv \pi/4 \in$ (a)

THE PROBLEM OF THE RESOLUTION

<< Given a plane triangle is said to be solved if, given three elements, one of which at least one side, there are the other three elements >>

It must be concluded that data items∈ (a), that the only three corners, the triangle is not solvable given that there are infinitely many triangles and the like that have the same angles but different sides. Prevents this case is precisely the theorem of Pythagoras In fact, the solution is to miurare two sides b and c and if b = c, we have $2b^2 = a^2$ (1) from which it follows: $a= b\sqrt{2}$ (2).It remains to point out that the surveyous (**Pythagoras**) and geodetic (Eratosthenes) of the time, used the stadia (about 2.8 meters) to measure the dis-tanze and the sextant, **Fig 27-Strum**, to measure to measure angles. Used by **Hipparchus** the Earth-Luna. In case (1) the Pythagorean school, Metaponto, was able to draw on the ground of trinaoli rectangles and then to construct equilateral triangles (a) any rectangles and triangles (b) . The difference in the resolution is crucial. In fact, (a) is solved with the arithmetic (2) and (b) requires trigonometric operators In fact, suppose you know a = 120 m, $\beta = 38^o28$ '. Then by Theorem of Pythagoras we can to write

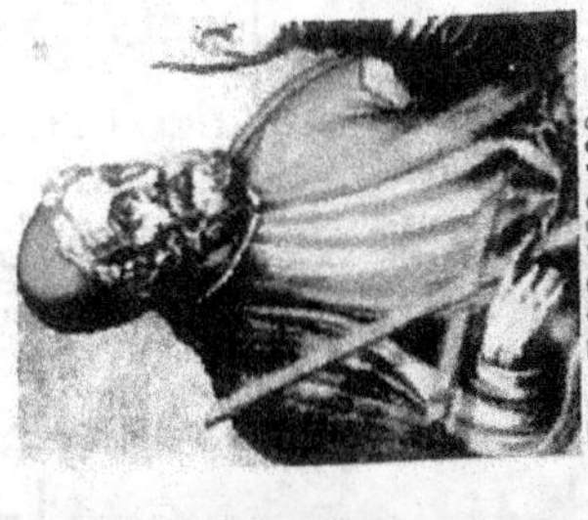
IPPARCO 190-120 a.c.

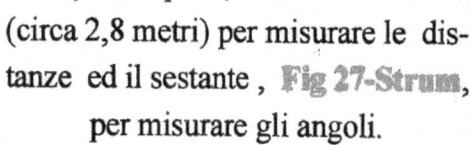

COMPLETAMENTO RISOLUZIONE | Fig-5

Nel caso (b) abbiamo i dati a = 120 m, $\beta = 38^o 28'$ e poichè il triangolo è rettangolo si può scrivere :

1- **Risoluzione angolare** del triangolo A-B-C:
$\alpha+\beta+\gamma=180^o \rightarrow \gamma = 180^o-90-38^o28' = 40^o\ 07'\ 30''$

2- **Risoluzione dei lati b** e **c** . Si ottiene :
b=a sin $38^o10'$ = 120. 0,6170358= 74,044 m ,
c =120.sin $40^o07'30''$=120.0,643783=77,25 m .

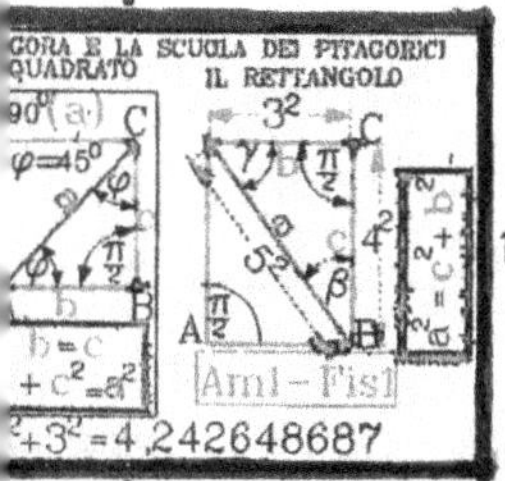

Torniamo al secolo X dopo C. Quando l'Europa in lettargo scientifico a partire , secondo nostra suddivisione storica . Dalla morte di **Archimede** (212 d.C.)e la dominazione **S.P.Q.R**.(**Senatus populosque romanus**) Gli Arabi hanno occupato due dei tre continenti dopo le macerie di Roma maestra del potere (che logora chi non ce l'ha) ma ne distrugge l'anima mandando Cristo a morire tra i ladroni. Gli Arabi si sono presentati con la scimitarra sguainata ma non dobbiamo dimenticare che hanno conservato, diffuso, inventato il circulum parvulus, cioè le basi della aritmetica . In breve hanno inventato la aritmetica ciffrata del sistema digitale :Con ciò la polarità 10 Nord Sud(Zenit-Nadir)

Hanno raccolto quanto di scientifico proveniente dalla Cina e dall'India. Sulle orme della scienza speculativa dei Greci hanno conservato parte degli elementi di **Euclide** con degli interessanti sviluppi di algebra .di **Abu'l Fath Muhhamad (ibn figlio) Quasim** sulle coniche di **Apollonio** con interessanti deduzioni . Uno dei maggiori esponenti della cultura Araba,**Abu' Abdallah Muhammed ibn Gabir ibn Sinan Al butani** con commenti ed inserzioni originali sulle funzioni di trigonometria sferica,**Fig23-Astr1**. Riportiamo salvo approfondimento a seguire le seguenti relazioni:

$\sin b=\frac{\sin c}{\sin C}$ (1) , $\sin \alpha=\frac{\tan c}{\tan C}$ (2) , $\cos A=\frac{\tan c}{\tan C}$ (3)

ed altre relazioni , a= numeratore di un angolo del piano e A, B angoli sferici di **Muhammed ibn Muhammed ibn Jhaia ibn Imail ibn Wafa**

COMPLETION OF RESOLUTION | Fig-5

In case (b) we have the data a = 120 m, $\beta = 38^o 28'$ and because the Triangle is right you can write:

1 - **Angular resolution** of the triangle A-B-C:
$\alpha+\beta+\gamma = 180^o \rightarrow \gamma = 180^o -90-38^o28' = 40^o\ 07'\ 30''$

2 - **Resolution of the sides b** and **c**. You get:
b = a sin $38^o10'$ 00= 120. 0.6170358, m = 74.044,
c =120.sin $40^o07'30''$ = 120.0,643783, m = 77.25.

Let's go back to the tenth century after C. When Europe in lettargo Scientific to leave, according to our south-historical division.

Since the death of **Archimedes** (212 C.d.) and the domination SPQR (Senatus Romanus populosque) The Arabs have occupied two of the three continents after the rubble of Rome mistress of power (which wears out those who do not have it) but it destroys' soul by sending Christ to die between the two thieves. The Arabs are presented with the sword unsheathed but we must not forget that they have preserved, disseminated, invented the circulum parvulus, that is the basics of arithmetic. In short they invented arithmetic ciffrata of the digital system. Thereby the polarity 10 North South (Zenit-Nadir)

They collected as of scientific coming from China and India. In the footsteps of the speculative science of the Greeks have preserved some of the elements of **Euclid** with the interesting developments of algebra. **Muhhamad of Abu'l Fath (ibn son)** of conics of **Apollonius** Qasim with interesting deductions. One of the greatest exponents of Arab culture, **Abu 'Abdallah ibn Muhammad ibn Sinan Al Gabir butanes** with comments and posts on the functions of the original spherical trigonometric-tria, **Fig23-Astr1**. Here except deepening to follow the following relations:

$\sin b=\frac{\sin c}{\sin C}$ (1) , $\sin \alpha=\frac{\tan c}{\tan C}$ (2) , $\cos A=\frac{\tan c}{\tan C}$ (3)

reports, a = numerator of a plane angle and A, B spherical corners of **Muhammed ibn Muhammed ibn ibn Jhaia Imail ibn Wafa.**

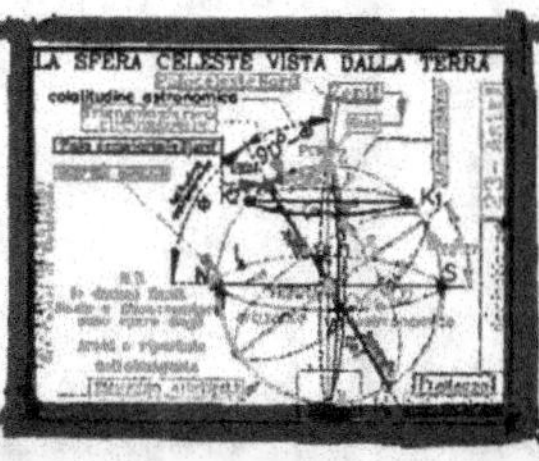

ARCHIMEDE 287-121a.c.

LA ASTROFISICA DA BABILONIA Fig-6

Nasir ed Din nato a Tus nel Febbraio 1201(Corassan) e vissuto riporta le formule di trigonometria sferica che abbiamo promesso di illustrare in termini moderni riprese dai vari autori dell'epoca a cominciare da **ibn Gabir ibn Sinan Al butani** con commenti ed inserzioni originali sulle funzioni di trigonometria sferica :

$\sin b = \frac{\sin c}{\sin C}$ (1) , $\sin \alpha = \frac{tn\, c}{tn\, C}$ (2) $\cos A = \frac{tn\, c}{tn\, C}$ (3)

Quale valore probatorio hammo queste formule ? La risposta richiede una approfondita ricognizione delle grandezze in gioco e sui mezzi per la misura.

LA FISICA ED IN MOTI DEGLI ASTRI

Abbiamo appreso dalla fisica classica e dalla meccanica che un corpo possiedono un propio moto rettilineo uniforme(Galilei)se altre forze non intervengo a modificarne lo stato,

P1- I moti della Terra . Se ne contanto almeno 15 I 4 principali sono : rivoluzione kepleriana attorno al Sole,rotazione attorno al suo asse polare (PCN) rotazione attorno all'asse equatoriale(PEN)detto di precessione.Infine di nutazione.

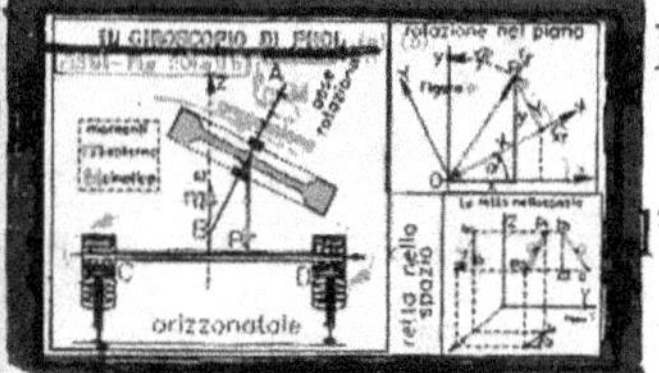

La Fig- Fisb1 . Come mostra la simulazione giroscopica l'effetto trottola innesta un moto di nutazione mentre quello di precessione tende a stabilizzare il corpo. Allo stesso modo la Terra possiede appunto i quattro moti descritti . La nutazione dal punto di vista fisico,Fisb1-Fig21,è realizzata con il

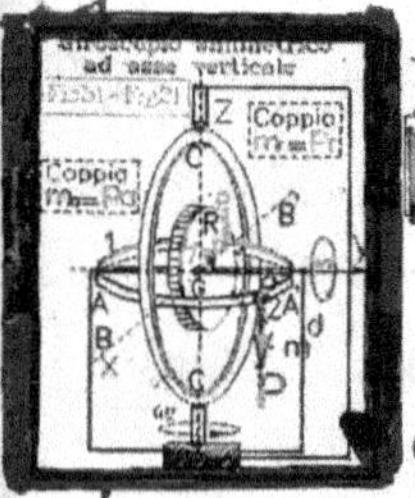

giroscopio Infatti applicando una coppia $P . \bar{y} = m_x$ la energia si trasferisce della ruota R per effetto della coppia corrispondente che ruota alla pulsazione ω_y di energia cinetica Wc. Il Fisico Phol ha in tal modo dimostrato che è possibile trasferire energia cinetica in energia di rotazione. La Terra, il cui asse di rotazione (PCN) è diretto verso la stella polare è inclinato sull'asse di rotazione terrestre

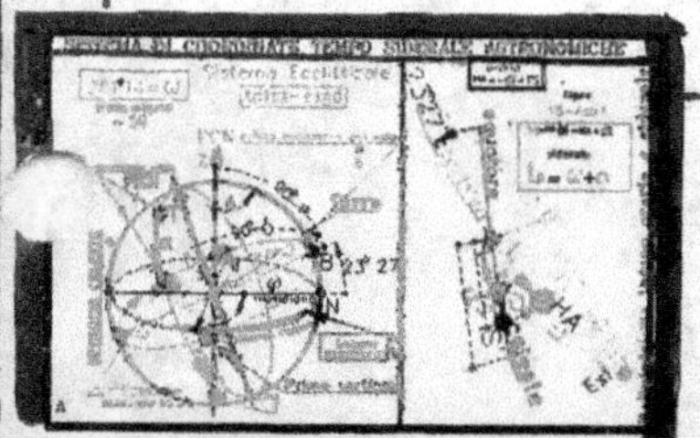

THE ASTROPHYSICS OF BABYLON Fig 6-

Nasir ed-Din was born in Tus in February 1201 (Corassan) and lived shows the formulas of spherical trigonometry we have promised to explain in modern terms of the time taken by the various authors starting with **Gabir ibn Sinan ibn Al butanes** with comments and posts on the functions of the original spherical trigonometry::

$\sin b = \frac{\sin c}{\sin C}$ (1) , $\sin \alpha = \frac{tn\, c}{tn\, C}$ (2) , $\cos A = \frac{tn\, c}{tn\, C}$ (3)

What probative value hamm these formulas? The answer requires a thorough reconnaissance of the magnitudes involved and the means for measurement.

PHYSICS AND MOTIONS OF THE STARS

We have learned from classical physics and mechanics that a body possess a propio uniform rectilinear motion (Galilei) if other forces did not come to inter-change the status,

P1-The motions of the Earth. If you contanto at least 15 The 4 main ones are: Keplerian revolution around the Sun, rotation around its polar axis (PCN) otation around the equatorial (PEN) said precetion Infine nutation.

The Fig-Fisb1. As shown in the simulation giroscopic effect spinning engages a motion of precession nutation while that tends to stabilize the body. Similarly the Earth has precisely the four motions described.

The nutation from the physical point of view, Fisb1-Fig21, is made with the gyro fact applying a torque moves $P . \bar{y} = m_x$ the energy of the wheel R for that wheel effect of the torque, corresponding to the pulsation wy of kinetic energy Wc. The Body Phol has thus shown that it is possible to transfer kinetic energy into rotational energy.

The Earth, whose axis of rotation (PCN) is directed towards the polar star is tilted axis (PEN) of 23° 27 'Earth's axis ....

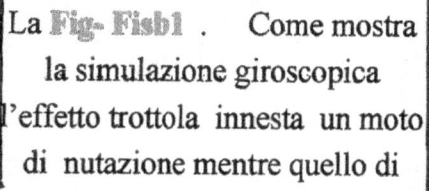

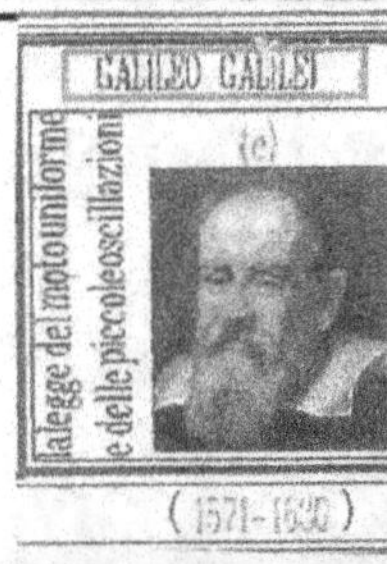

LA ASTROFISICA DA BABILONIA Fig-7

Nasir ed Din nato a Tus nel Febbraio 1201 (Corassan) e vissuto riporta le formule di trigonometria sferica che abbiamo promesso di illustrare in termini moderni riprese dai vari autori dell'epoca a cominciare da **ibn Gabir ibn Sinan Al batani** con commenti e con delle ~~inserzioni originali sulle funzioni di trigonome~~-tria sferica del tipo visto $\sin\beta=\frac{\sin c}{\sin C}$ (1), $\sin\alpha=\frac{\operatorname{tn} c}{\operatorname{tn} C}$ (2), $\cos A=\frac{\operatorname{tn} c}{\operatorname{tn} C}$ (3)

Quale valore hammo queste formule ? Da un punto di vista storico rilevante, Esse provano che in antichità la astronomia era tenuta in gra conto Non veniva coltivata ma proprio gli Arabi sono stati la memoria storica dell'atrionomo Tolomeo che dominò le coscienze con la sua ipotesi della creazione dell'universo subordinato dalla Terra **Copernicò** demolì il sogno dell'uomo di collo-carsi(inter pares)con la creazione e **Keplero** deduce le leggi che la suborina collocandola nel giusto posto assegnatogli da **CHI** può ciò che vuole .Ma quando si passa dalla osservazione degli astri alla astronomia di posizione si dovrà attentendere **Eulero**,, **Briggs** ,ed altri per dare alla trigonometria sferica un modello funzionale per la conoscenza del cielo e delle stelle .

La **fig 2-Astr2** di fondo pagina .

Segnaliamo perciò le caratteristiche che il sup-porto grafico permette di dedurre.

1— Come si può osservale gli archi **a,b,c** sono dei lati di triangoli sferici quindi essi sono pienamente definiti quando si conosca il raggio R della sfera celeste di appartenenza.

2- La misura corrispondente $\frac{a}{R}, \frac{b}{R}, \frac{c}{R}$ se si pone R=1 si trasformano in angoli di un trian-golo piano,cioè $\lim_{R\to 1}\frac{a}{R}=\alpha$ (4) e cosi per β,γ

THE ASTROPHYSICS OF BABYLON Fig 7

-**Nasir ed-Din** was born in Tus in February 1201 (Corassan) and lived shows the formulas of spherical trigonometry we have promised to explain in modern terms of the time taken by the various authors starting with **Gabir ibn Sinan ibn Al butanes** with comments and with the original ad on triaspherical trigonometric ~~functions of the type seen~~ : $\sin\beta=\sin b=\frac{\sin}{\sin}$ (1), $\sin\alpha=\frac{\operatorname{tn} c}{\operatorname{tn} C}$ (2), $\cos A=\frac{\operatorname{tn} c}{\operatorname{tn} C}$ (3)

What value hamm these formulas? From a historical point of view relevant They prove that in ancient astronomy was held in the gra account was not cultivated but just the Arabs were the historical memory dell'atrionomo Ptolemy who ruled the conscience with his hypothesis of the creation of the universe subject from Earth **Copernicus** demolished man's dream of neck-carsi (peer) with the creation and **Kepler** to deduce the laws that suborina placing it in the right place assigned to him by **WHO** can be what he wants. But when it comes from the observation of stars to astronomy position you will have attentendere **L.Euler**(1707-1783), **E. Briggs** (1556-1630 , and others to give to spherical trigonometry a functional model for the knowledge of the sky and stars. **Figure 2-Astr2** of the footer therefore point out the features that sup-port graph allows us to deduce.

1 - As you can observe them arcs **a, b, c** are the sides of spherical triangles so they are fully defined when you know the raggli R of the celestial sphere of belonging.

2 - The corresponding measure $\frac{a}{R}, \frac{b}{R}, \frac{c}{R}$, yes, if you put R = 1 are transformed into angles of a plane triangle, that is,

$$\lim_{R\to 1}\frac{a}{R}=\alpha \quad (4)$$ and so for β, γ

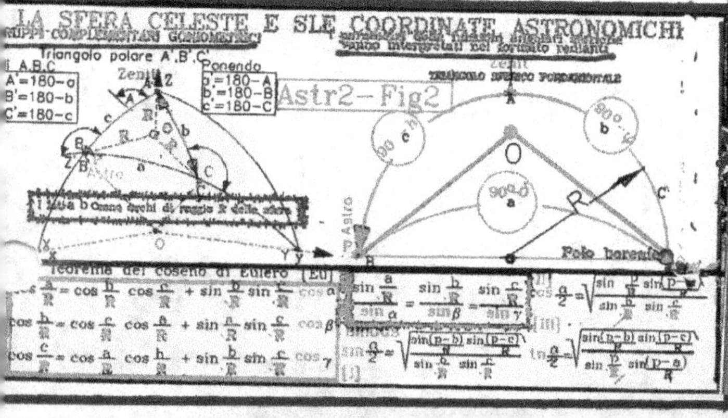

3- Keplero

COPERNICO 1473-1543

BRIGGS

IL POSIZIONAMENTO DEGLI ASTRI | Fig-8

Abbiamo visto le modalità con le quali si misurano gli archi su una superficie sferica di raggio R variabile fino all'astro o centrato sulla regione del cielo che interessa. Se l'oggetto è una data stella , Fig2-Astr2 , per la posizione è necessario scegliere un riferimento fisso rispetto alla Terra . **Copernico** ci assicura " Sol Sstat "

Gli astronomi hanno scelto vari sistemi di ccordinate celesti. Sorgono ora due problemi

1°- Come misurare le posizioni nella volta popolata di astri e come rappresntare lo sferoide?

2°- Quali mezzi sono necessari per lo scopo ?

Il rapido avanzare della tecnologia permette di rilevare la distanza degli astri: $d = zc/H_0$ (1), con **d** dell'ordine dei miliadi di anni luce !

Nel 300 (a.C.) **Ipparco** con l'uso del solo sestante misurava la paralasse lunare di 45' 15'' e , usando il raggio dello sferoide terrestre R≅ 7.000 km di Eratostene(R≅ 7.000 km)misurò la distanza d = distanza(Terra- Luna).≅ 400.000 km Oggi con il Radar 384.403 km. I grndi passi tecnologici hanno consentito progressi e nei riferimenti storici un po di confusione. E' il caso delle andiche classi di luminosità delle stelle che **Pogson** ha convertito in magnitudini partendo dalla massima del SOLE con m=-17 contro la Sirio appena visibile ad occhio nudo : m=+0,2 . Tutte i miliardi di stelle con m< +0,2 sono invisibili per l'occhio con i telescopi Per la rappresentazione grafica del cielo stellato e del triangolo sferico fondamentale il modello dwg precedente ,modificato Fig3-Astr2. Questo colloca la Terra(**punto di osservazione**)nell'emisfero fra il polo boreale ed Australe,fra i parametri definiti :

1- a→ ascensione retta di un astro siulla sfera R

2- δ = declinazione della stella in P

Se il raggio R ≠ R , allora δ ≠ δ Coordinate sferice celesti con R Variabile da 1 ad ∞ succesione di sfere per l'astro Antarese nel piano di Gaus in TAV.13 A

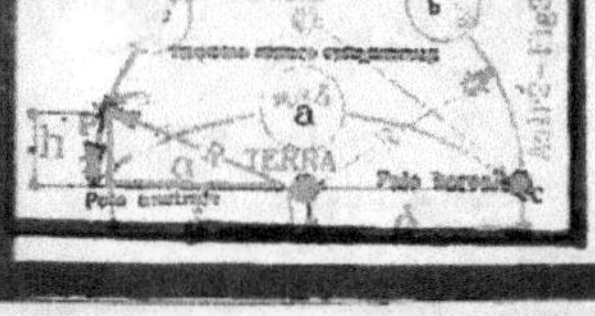

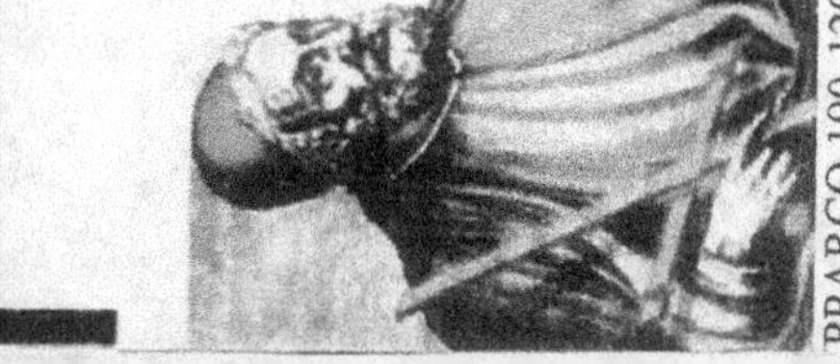

IPPARCO 190-120 a

THE POSITIONING OF THE STARS | Fig-8

We have seen how the qualui is me-surano bows on a spherical surface of radius R centered on the variable up to the star or region of the sky that interests you. If the object is a given star, Fig2-Astr2, for positioning, it is necessary to choose a reference at rest with respect to the Earth. **Copernicus** assures us that'' Sol Stat! ''

Astronomers have chosen various systems coordinate heavenly problems that arise now summarize:

1°-How to measure populated positions in the vault of stars? How rappresntare spheroid?

2° - What resources are required for the purpose?

The rapid advance of technology allows you to measure the distance of the stars: $d = zc / H_0$ (1), with d the order of bilion light years! In 300 (c.C) **Hipparchus** with the use of only ses-standing measured the lunar parallax of 45 'and 15'', using the radius of the Earth's spheroid R ≅ 7000 km of **Eratosthenes** (R ≅7000 km) measured the distance d = distance (*Earth*-Moon). with 400,000 km .Today 384,403 km with Radar. The grndi steps in technology have enabled progress in historical references and a bit of confusion. It 's the case of andic luminosity classes of the stars that **Pogson** converted in magnitudes from the max SUN with m = -17 against Sirius barely visible to the naked eye: m =+ 0.2. All the billions of star with m <+0.2 are invited to the eye hissing with telescopes, For the graphical representation of the starry sky and the spherical triangle fundamental dwg previous model, modified Fig3-Astr2.

This places the Earth (**observation point**) in the northern hemisphere between the pole and Southern Africa, between the parameters defined:

1 - α right ascension of a celestial sphera for R

2 - δ declination of the star P If the radius R≠ R, then δ≠ δ sferice celestial coordinates with R variable from 1 to ∞ succession of balls for the star Antares represented in the plan Gaus in TAV.13

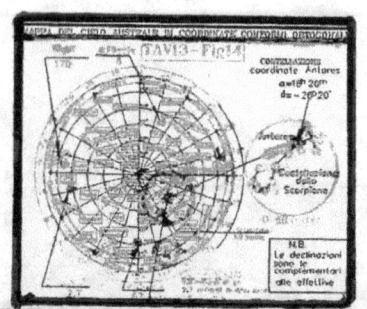

COPERNICO 1473-154

LE STELLE PIU' LUMINOSE — Fig-9

Abbiamo visto le modalità con le qualui si misurano gli archi su una superficie sferica di raggio R variabile fino all'astro o centrato sulla regione del cielo che interessa. Se l'oggetto è una data stella , Fig2-Astr1. Ma sono noti molti parametri che riportiamo nel CAD-12 con una breve descrizione dei paramentri delle stesse ,

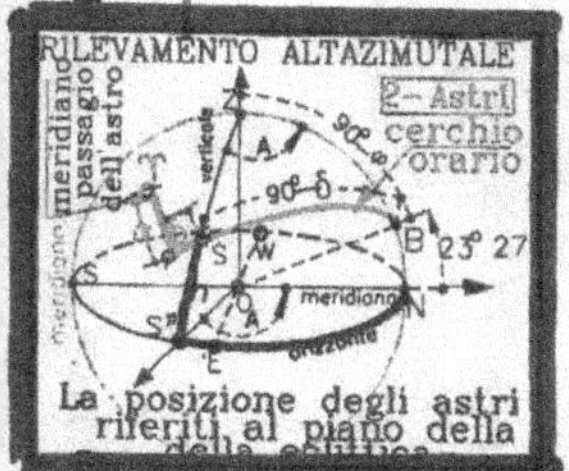

THE STARS MORE 'LIGHT — Fig-9

We have seen how the qualui is me-surano bows on a spherical surface of radius R centered on the variable up to the star or region of the sky that interests you. If the object is a given star, Fig2-Astr1

But many are known parameters that we report in CAD-12 with a brief description of the parameters of the same

TAB-A Tabulato

CARATTERISICA DELLE STELLE PIU' LUMINOSE CAD-12

N	Nome e Costellazioni	α	δ	m (Magnitudine)	Tsp (Classe spettrale)	p'' (Parallasse)	r (Distanza in parsec)	u (Velocità orbitale dell'Astro)
1	Achernar (α Eridani)	$1^h\ 36^m$	$-57^\circ\ 30'$	0,60	B_5	0'',045	21	19
2	Polare (α Ursae minoris)	1 47	+89 01	2,12	F_8	0'',012	85	variabile
3	Algenib (α Persei)	3 21	+49 41	1,90	F_5	0'',020	50	………
4	Aldebaran (α Tauri)	4 33	+16 25	1,06	K_5	0'',057	18	54
5	Rigel (β Orionis)	5 12	− 8 15	0,34	B_8	0'',006	170	variabile
6	Capella (α Aurigae)	5 13	45 57	0.21	G_0	0'',068	15	29
7	Tauri (β)	5 23	+28 34	1,78	B_8	0'',035	28	10
8	Belatrice (γ Orionis)	5 23	+ 6 18	1,70	B_2	0'',017	59	16
9	ε (Orionis)	5 34	− 1 14	1,75	B_0	0'',008	125	10
10	ξ (Orionis)	5 37	− 2 00	1,05	B_0	0'',008	125	19
11	Betelgeuse (α Orionis)	5 52	+ 7 24	0,50	Ma	0'',012	85	21
12	β (Aurigae)	5 56	+44 57	2,07	A_0	0''029	35	−10
13	β (Canis Maioris)	6 21	−17 56	1,99	B_1	0'',012	85	33
14	Canopo (α Navis)	6 23	−52 49	−0.86	Fb	0''016	65	20
15	γ (Geminonorum)	6 35	+16 26	1,93	A_0	0'',047	21	− 11
16	Sirio (α Canis Maioris)	6 43	−16 39	− 1,58	A_0	0'',375	2,7	− 7
17	ε (Canis Maioris)	6 57	−28 54	1,63	B_1	0'',012	85	28
18	δ (Canis Maioris)	7 06	−26 19	1,98	F_8	0'',010	100	
19	Castore (α Geminonorum)	7 31	+ 32 00	1.58	A_0	0''074	14	6
20	Procione (α Canisminoris)	7 37	+ 5 21	0.48	F_5	0'',310	3	− 3
21	Polluce (β Geminonorum)	7 42	+ 28 09	1,21	K_0	0'',110	9	4
22	γ (Navis)	8 8	− 47 11	1,92	Oa	………	………	35

Consider the star Rigel (α Orionis)

This **star** * celestial coordinate defined: $\alpha = 5^h12^m$ (right ascension), $\delta = -8^\circ16'$ (declination)

m = 0.84 **(magnitude Pogson)**,

TSP (termospettroB8) ,

Parallax = 0.009'',
The ecliptic distant r = 170 pc at 559.3 a.l

Consideriamo la stella Rigel (α Orionis)

Questa **stella** * di coordinate celesti definite: $\alpha=5^h12^m$**(ascensione retta)**,$\delta=-8^\circ16'$**(declina zione)** m =0,84 **(magnitudine Pogson)** , **TSP(termospettroB8) , Parallasse =0,009 ''** , **distanza r dalla eclittica 170 pc=559,3 a.l**

La stella Rigel (Tav2) è da considerarsi il limte nella sfera celeste degli oggetti luminosi risolvibili con i mezzi di osservazione dei cannocchiali astronomici.Infatti occorre che l'obbiettivo sia in grado di misurare la parallasse **=0,009 ''** del satellite della Rigel Mostriamo,salvo ritornare sopra il cannochiali alt-azimutale di Monte Mario (Roma)

The star Rigel (Tav2) is considered the limt in the celestial sphere of bright objects resol-vibili with the means of observation of annocchi-wings astronomici.Infatti necessary that the lens is capable of miurare the parallax = 0.009'' of the satellite We show the Rigel, except back over the Rifle alt-azimuth of Monte Mario (Rome)..

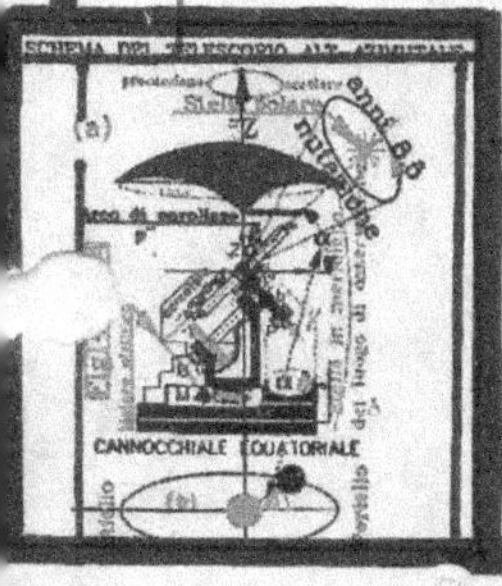

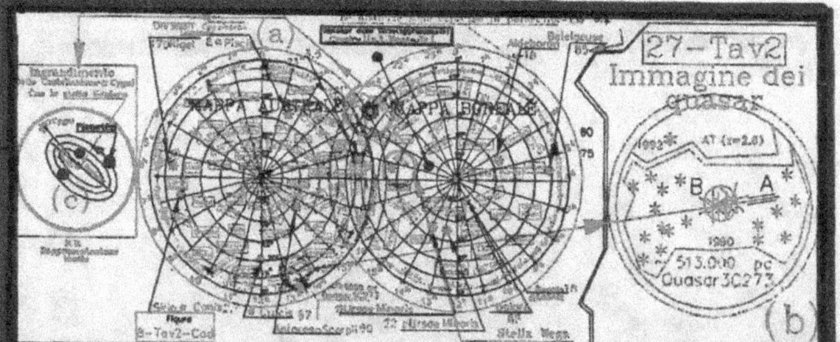

TAB-A Tabulato A

CARATTERISICA DELLE STELLE PIU' LUMINOSE CAD-12

N	Nome e Costellazioni	α	δ	m (Magnitudine)	Tsp (Classe spettrale)	p''	r (Distanza in parsec)	u (Velocità orbitale dell'Astro)
1	Achernar (α Eridani)	1^h 36^m	$-57°$ 30'	0.60	B_5	0'',045	21	19
2	Polare (α Ursae minoris)	1 47	+89 01	2,12	F_8	0'',012	85	variabile
3	Algenib (α Persei)	3 21	+49 41	1,90	F_5	0'',020	50	....
4	Aldebaran (α Tauri)	4 33	+16 25	1,06	K_5	0'',057	18	54
5	Rigel (β Orionis)	5 12	− 8 15	0,34	B_8	0'',006	170	variabile
6	Capella (α Aurigae)	5 13	45 57	0,21	G_0	0'',068	15	29
7	Tauri (β)	5 23	+28 34	1,78	B_8	0'',035	28	10
8	Belatrice (γ Orionis)	5 23	+ 6 18	1,70	B_2	0'',017	59	16
9	ε (Orionis)	5 34	− 1 14	1,75	B_0	0'',008	125	10
10	ξ (Orionis)	5 37	− 2 00	1,05	B_0	0'',008	125	19
11	Betelgeuse (α Orionis)	5 52	+ 7 24	0,50	Ma	0'',012	85	21
12	β (Aurigae)	5 56	+44 57	2,07	A_0	0''029	35	−10
13	β (Canis Maioris)	6 21	−17 56	1,99	B_1	0'',012	85	33
14	Canopo (α Navis)	6 23	−52 49	−0,86	F_0	0''016	65	20
15	γ (Geminonorum)	6 35	+16 26	1,93	A_0	0'',047	21	− 11
16	Sirio (α Canis Maioris)	6 43	−16 39	− 1,58	A_0	0'',375	2,7	− 7
17	ε (Canis Maioris)	6 57	−28 54	1,63	B_1	0'',012	85	28
18	δ (Canis Maioris)	7 06	−26 19	1,98	F_8	0'',010	100	
19	Castore (α Geminonorum)	7 31	+32 00	1,58	A_0	0''074	14	6
20	Procione (α Canis minoris)	7 37	+ 5 21	0,48	F_5	0'',310	3	− 3
21	Polluce (β Geminonorum)	7 42	+ 28 09	1,21	K_0	0'',110	9	4
22	γ (Navis)	8 8	− 47 11	1,92	Oa	....	....	35

UNIVERSO STELLARE ELIOCENTRICO VISTO DALLA TERRA

Fig14 – Astr1

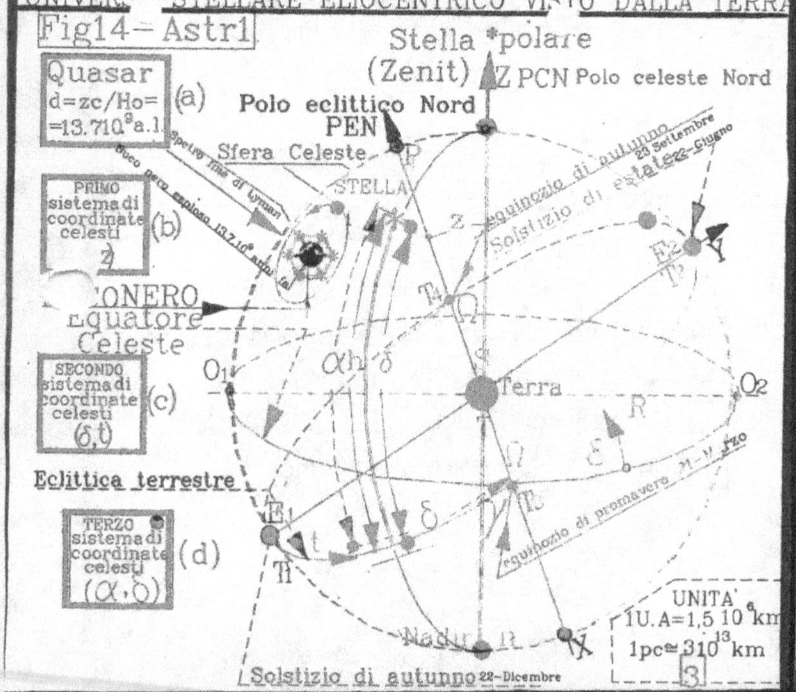

. La m è la magnitudine o emittanza della sorgente Si noti che il Sole ha m=-27 watt è la stella per la Terra di massima magnitudine secondo Pogson . La stella Sirio è appena visibile ad occhi nudo, con m= 0 Per valodi m>0 la stella è invisibile sensa l'apporto di adatti equipaggi ottici

In antichità le stelle erandate in classi

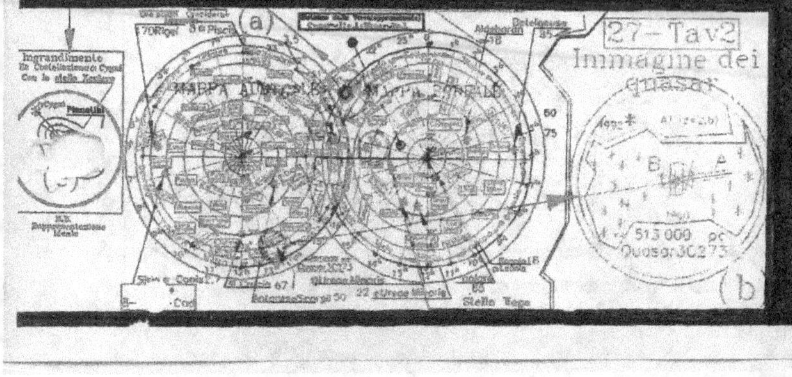

LE MAGNITUDINI STELLARI DI POGSON

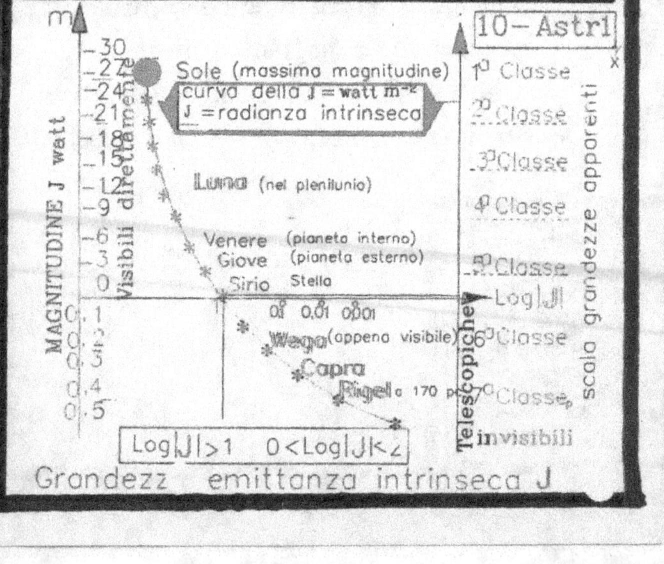

Ricordiamo brevemente le tre leggi dell'ottica geometrica riferita ai raggi luminosi in un mezzo materiale , riflettente e/o rifrangente , caratteriz-zata da alcune proprietà regolate dalla legge di **Villebrord Snell Van Royen** , latinizzato con **Snellius**(1580-1626) detto Snell.

Fisico,matematico , geodeta Olandese autore del teorema in concorso con Pothemot sulla

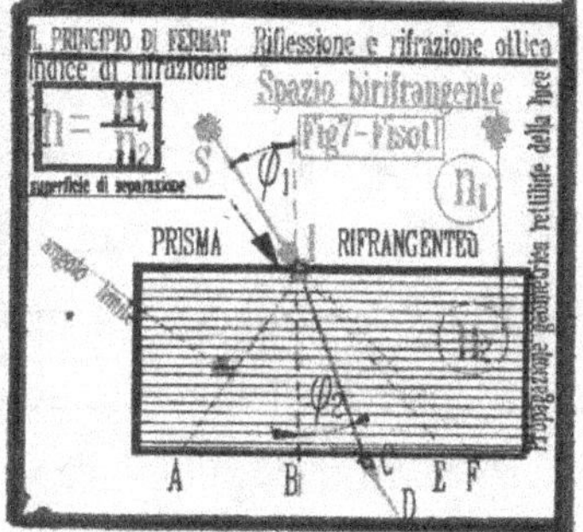

risoluzione di reti to-pografiche in applicazioni di trigonometria piana e sferica. In tali applicazioni si realizzano sistemi prismatici per il tracciamento di allineamenti rettilinei. Ciò dà la misura del principio di **Fermat** di **Fig-7** Un raggio di luce, che parte dalla sorgente **S** nel mezzo **n**1 e incide sul prisma in **I** , per ci passa na normale con un angolo φ_1. Se la il prisma non è argentato e se $\varphi_1 > 0$ il raggio proseguirà il suo cammino nel mezzo **n**2, avvicinandosi alla normale di un angolo $\varphi_2 < \varphi_1$

La legge di **Snell** afferma che il rapporto fra gli angoli dipende solo dagli indici di rifrazione dei mezzi attraversati. Postulata dalla equazione:

$$n_{12} = \frac{n_1}{n_2} = \frac{\sin\varphi_1}{\sin\varphi_2} \quad (1) \quad \text{con} \quad n_{12} = \text{Costante} \ (1_1).$$

Allora se si pone $\varphi 1 = \pi/2$ si trova la condizione :

$$\sin\varphi 2 = \frac{n_2}{n_1} = n_{21} \quad (2) \quad \text{con } n_{21} = \text{Costante} \ (2_1)$$

La(2) prova la invertibilità dei raggi di luce

Se la sorgente **S** è posta in **n**2 la luce si riflette in **I** nello spazio fra i **raggi** per A e F come se la luce si riflette,**Fig3** , sulla superficie argentata

Let us briefly recall the three laws of geometrical optics referring to the rays of light in a material medium, reflective and / orifrangente, character-ized by some property under the laws of **Villebrord Snell Van Royen**, Latinized with **Snellius** (1580-1626) **Snell** said.

Physicist, mathematician, surveyor Dutch author of the theorem in competition with the Pothemot

to-polygraphic resolution of networks in applications of plane and spherical trigonometry. In such applications are realized prismatic systems for the tracking of alignments straights. This gives the measure of Fermat's principle in **Fig-7**

A ray of light, which starts from the source S in the middle **n**1 and affect the prism **I**, passes na normal at angle φ_1.

If the prism is silver and if $\varphi 1 > 0$, the beam will continue its journey in the middle n2, approaching the normal angle $\varphi 2 < \varphi 1$

The **Snell's** law states that the relationship between the angles of refraction depends only dagliindici of the means through which . Postulated by the equation:

$$\mathbf{n}_{12} = \frac{n_1}{n_2} = \frac{\sin\varphi_1}{\sin\varphi_2} \quad (1) \text{ with } \mathbf{n}_{12} = \text{Constant} \ (1_1).$$

So if you put $\varphi 1 = \pi / 2$ is the condition:

$$\sin\varphi 2 = \frac{n_2}{n_1} = n_{21} \quad (2) \quad \text{con } n_{21} = \text{Costante} \ (2_1)$$

The (2) test the reversibility of light rays

If the source **S** is placed in the light **n**2 is reflected in the space between the **rays** A and F as if the light is reflected, **Fig3**, on the silver surf

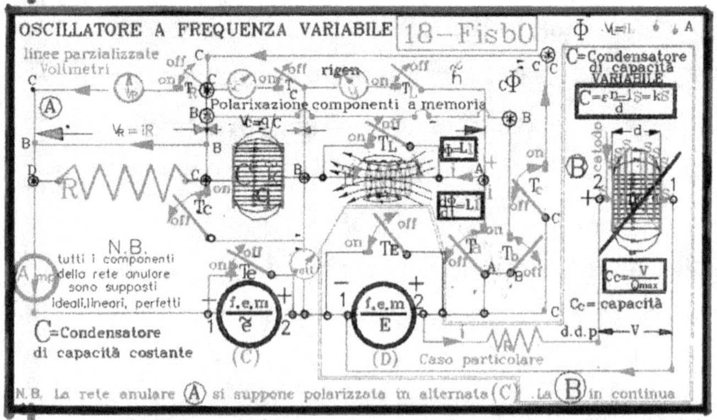

[I]Descrizione sommaria dei componenti

Interruttori T. Te nella posizione off invia al dispositivo $\tilde{Z}$(RCL) il segnale,se tutti i restanti T sono in on(cortocircuitazioneC C)e si polarizza: $\tilde{Z} = R + j\,(L\omega - \frac{1}{C\omega})$ (a) la impedenza $\tilde{Z}$ Allo scopo basta porre Ta in on. Al solito il simbolo $\tilde{e}(t) = Em \sin(\omega t + \phi)$ (1C) ,con Em ampiezza della f.e.m., $\omega = 2\pi f$ pulsazione di frequenza propria f,fase dell'ingresso(1).Chiudere od aprire un un interuttore **T**significa trasformare topologicamente il dispostivo funzionante. I componenti (R=reistore,**C=condensatore,L=induttore**) sono supposti ideali significando che tanta corrente entra da un morsetto altrettanta ne esce. Quanto alle caratteristiche elettriche la R dissipa energia mentre L e C sono a memoria e conservano la energia del campo {$\underline{k}$ elettrico} e {$\underline{h}$ magnetico}

Il problema del funzionamento.

[I] La impedenza $\tilde{Z}$ polarizzata dalla $\tilde{e}(t)$

Se all'istante $t_0=0$ si chiude Ta, con T_E in on **E(t)** soppressa allora la $\tilde{Z}$ è polarizzata e quindi nel circuito circola una corrente di indotta dalla f.e.m. (1C) . Supposto che i componenti a memoria siano nello stato zero cioè scarichi la condizione di equibrio tensio di ingresso e corrente in uscita è ,in termini di tensione applicata (1C) e le cadute di potenziali ad opera di ogni singolo componente così definita in tempo reale:

$$V_L(t) + V_C(t) + V_R(t) = Em \sin(\omega t + \phi) \quad [1]$$

[I] Brief description of components T.

Te switches in the off position sends the device signa $\tilde{Z}$(RCL) if all the remaining T are on (to short circui C C) and polarization of: $\tilde{Z} = R + j\,(L\omega - \frac{1}{C\omega})$ (a) in order impedance just put in on Ta . Usual symbol : $\tilde{e}(t) = Em \sin(\omega t + \phi)$ (1C), with Em amplitude of the **e.m.f**, $\omega = 2\pi f$ pulsation of frequency intrinsic f , phas of the input (1) . Close or open a switch a **T** significa transform the mind topological dispostivo working.

he components (R = reistore , **C = capacitor** , **L = inductor**) are supposed ideals to signify that much current to go from a terminal just as it comes out . As for the electrical characteristics of the R dissipates energy while L and C are in memory and preserve the energy of the electric field {$\underline{k}$ } and {$\underline{h}$ }

The problem of magnetic operation.

[I] The impedance $\tilde{Z}$ polarized by $\tilde{e}(t)$

If the instant $t_0 = 0$ see closes Ta , with T_E on **E (t)** is deleted then the polarized and then in the circuit circula motion of current induced **e.m.f** (1C).

Assumed that the components in memory are in the zer state that discharges the condition of equibrio voltage input and current output is , in terms of the applied voltage (1C) and the falls of potential to work of each component single so defined time real :

$$V_L(t) + V_C(t) + V_R(t) = Em \sin(\omega t + \phi) \quad [1]$$

[I]La corrente in uscita Fig 24-Fisb0

Nel caso della maglia **(a)**, equivalente a quella dell'oscillatore semplificata con il solo ingresso in alternata: $\tilde{e}(t)=E_m\sin(\omega t)$ (1).Si supponga che i

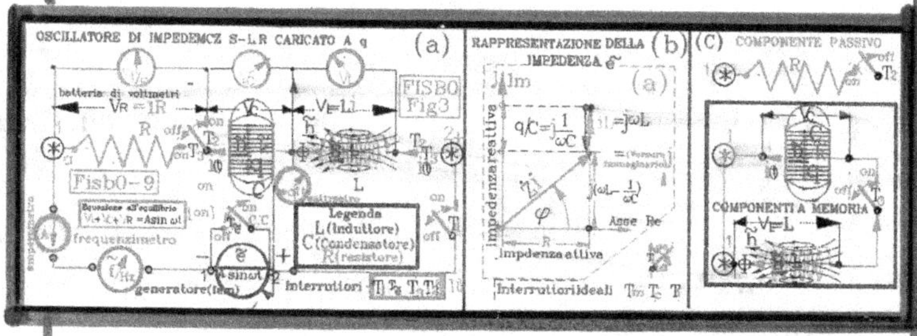

componenti a memoria (C ed L) siano scarichi cioè q=0 Φ=0 all'istante t_o=0 . Instante nel quale si chiude T_1.Questa chiusura(T_1in on) ha l'effetto di polarizare la impedenza Ż ,nicchia (b) .Si ottiene in tal modo una corrente in uscita relativa alla **f.e.m.** (1). L'equilibrio istantaneo fra le cadute di potenziale dei componenti e tensione di ingresso risulta,per t>0 in tempo reale,retto dalla relazione:

$$V_L(t)+V_C(t)+V_R(t)=E_m\sin(\omega t) \quad [1]$$

[II] La corrente di uscita della [1]

Dalla teoria delle reti elettriche nel caso della **[1]** la soluzione generale è data dalla equazione :

$y(t)=y_o(t)+ y_p(t)$ (2), da cui l'enunciato :

<La uscita in corrente di una impedenza Ż polarizzata da una f.e.m. sinusoidale è uguale alla somma della sua omogenea associata e da una sua equazione particolare dipenedente dai parametri K_1 e K_2 dei parametri iniziali>

[III]Soluzione in corrente della omogenea $y_o(t)$

La omogene associata della [1]si ottiene eguagliando a zero la stessa,cioè :

$y_o(t)=V_L(t)+V_C(t)+V_R(t)= 0$ (2) .A questo punto occorre ricordare che le componenti risultano , in termini di caduta del potenziale ,della forma:

$V_L(t)= \frac{d\Phi^2}{dt^2}=Li'\frac{d\Phi}{dt}=Li''$ (2_1), $V_C(t)=\frac{dq}{dt}/C=i/C$ (2_2)

$V_R(t)=Ri'$ (2_3),che sostituite nella (2) porgono:

$Li''+Ri'+Ci=0$ (3), oppure : $i''+\frac{R}{L}i'+i/CL=0$ (4)

[I]The output current Fig24 Fisb0 In the case mesh **(a)**, equivalent simplified to that of the oscillator with the input only in : $\tilde{e}(t)=E_m\sin(\omega t)$ (1) .

Suppose that the components in memory (C and L) have been discharged ie q = 0 to Φ = 0 at time t_o= 0. Instant in which it closes T_1.Questa closure (T_1 in on) has the effect of the impedance polarizare Ż, niche **(b)** Yes ing it in this way an output current on the **e.m.f** (1) The equilibrium between the instantaneous potential drops of the components and the input voltage is , for t> 0 in real time, the rectum by the relation:

$$V_L(t)+V_C(t)+V_R(t)=E_m\sin(\omega t) \quad [1]$$

[II] The courenti in output then [1]

from the theory of electrical networks in the case of [1], the general solution is given by the equation :

$y(t) = y_o(t) + y_p(t)$ (2) , by which the utterance :

< the current output Ż impedance of a polarized by an emf sine wave is equal to the sum of its associated homogeneous equation and a particular dependent its parameters K_1 and K_2 of the initial parameters >

[III] in the current solution of homogeneous $y_o(t)$ of the associated homogenous [1] is obtained equals - went to zero the same, namely :

$$y_o(t)=V_L(t)+V_C(t)+V_R(t)= 0 \quad (2).$$

At this point it should be recalled that the components are , in terms of fall of the potential , of the form following:

$$V_L(t)=\frac{d\Phi^2}{dt^2}= Li' \frac{d\Phi}{dt}=Li'' \quad (2_1)$$

$$V_C(t)=\frac{dq}{dt} C=i/C \quad (2_2), \quad V_R(t)=Ri' \quad (2_3)$$

which replaced in (2) gives :

$Li'' + Ri' + i/C = 0$ (3), too :

$$i''+\frac{R}{L}i'+i/CL=0 \quad (4)$$

COMPLEMENTO DESCRITIVO

Anticipiamo allo scopo di stabilire un contatto puramente indicativo fra la fisica e la elettronica dell'oscillatore RCL ad elementi ideali R - C - L

[I] Caso della depolarizzazione di Ż Infatti se **l'interruttore T_q** è in **on** la impedenza Ż ,alla chiusura di To non **impone corrente** i .

[II]Caso di carica intrappolata nel condensatore C Alla chiusura di To , poniamo in t_o =0

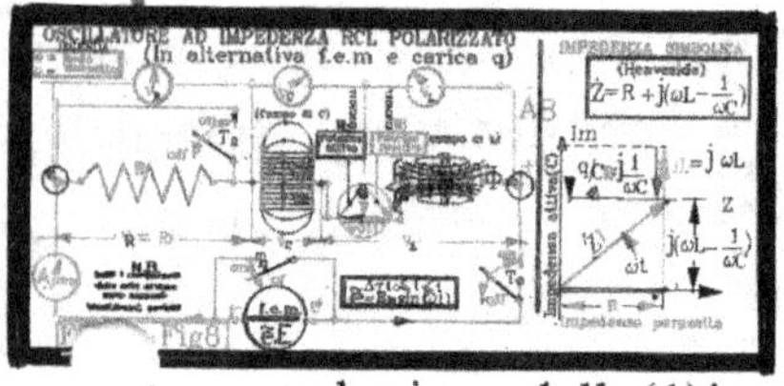

La cosidetta scarica del **C** (generatore potenziale) La equazione di equilibrio iniziale è: VC+VL+VR=0 (1). Come detto(pg.11) la soluzione della(1)implica,per l'uscita in corrente i ,la costruzione di una equazione differeenzale di ordine2 nella uscita con V come ingresso.

[III]Costruzione della equazione differenziale

I parametri risultano così definiti dalla relazione componenente-corrente.Precisamente:

{resistore:V_R=Ri (1_1), Induttore:V_L: Φ=Li $\frac{d\Phi}{dt}$=Li' (1_2) , condensatore: V_C= q/C (1_3) } .Queste relazioni introdotte nella (1) si scrivono:

q/C+Li'+Ri =0 (2) . Per ottenere la corrente occorre derivare la carica q, cioè dq/dt=i e di conseguenza anche le altre componenti. Ciò fatto risulta: i/C+Li''+ Ri' =0 (3) che ordinata è : i''+ pi'+q=0 (4) , con $p=\frac{R}{L}$ e $q=\frac{1}{LC}$. La teoria insegna che la (4) ha gli autovalori forniti dalla equazione caratteristica,cioè: $\lambda^2+p\lambda+q=0$ (5)

Dal**T.F.E(Teorema fondamenatele dell'algebra)** " se i p e q sono positivi è ideali la(5)ammette due radici complese coniugate. Risolvendo si ottiene: $\lambda_{1,2}=-p\pm j\sqrt{4q-p^2}$ (5') che soddisfano la (5) .

Si hanno 2 soluzioni particolari, Fig10d-Fis38 $i_1(t)=K_1[e^{-pt}e^{j\lambda 1t}]$ (6) , $i_2(t)=K_2[e^{-pt}e^{-j\lambda 2t}]$ (7)

Con le condizioni al limite e ponendo $K_1=K_2=I$ si trova $i(t)_{tot}=Ie^{-pt}\ [e^{j\lambda 1t}+e^{-j\lambda 2t}]/2$ (8) . Infine dalla[2]pg-48: $i(t)_{tot}=Ioe^{-pt}\ \cos(\omega_o t+\phi)$ (9) ,con un picco in t=0 e decrescrenza asintotica .La pulsazione $\omega_o=1/\sqrt{CL}$ ipende dai componenti C-L.

SUPPLEMENTARY DESCRIPTIO

We anticipate in order to establish contact between the indicative and the electronic oscillator physic RCL elements ideals R - C - L

[I] Case of depolarization Ż . In fact if the swich **Tq** in **on** the impedance , the closure of **To** not **impose currents** .

[II] Case of trapped charge in the condenser C

At the end of **To** , say at to = 0 The socalled discharge of **C** (generator potential) **.** The equilibrium equation is VC + VL + VR = 0 (1) As mentioned (pg.11) the solution of (1) implies AC , for the current output , the costrution of an equation of i second differenzial output if you order in the asssume V as input .

[III] The construction of the differential equazzione

The parameters are as defined by the relation componenente - current . precisely:

{Resistor: V_R = Ri (1_1) Inductor: VL : Φ=Li $\frac{d\Phi}{dt}$=Li' (1_2) capacitor : V_C = q / C (1_3)}.

These relations introduced in (1) allow you to write : q/C+Li'+Ri =0 (2)

To get the current necessary to derive the charge q , ie, dq/dt=i and of consequently also the other components . This fact results : i/C+Li''+ Ri' =0 (3) that is ordered i''+ pi'+q=0 (4) with $p=\frac{R}{L}$ and $q=\frac{1}{LC}$

The theory teaches that (4) has the eigenvalues provided by the characteristic equation , namely: $\lambda^2+p\lambda+q=0$ (5)

Dal**T.F.E (Theorem fondamenatele algebra)** "if i p and q are positive if the is ideal (5) admits two complex conjugate roots . Solving we get: $\lambda_{1,2}=-p\pm j\sqrt{4q-p^2}$ (5') , that satisfy the (5).

You have 2 special solutions , Fig10d - Fis38 ,ie: $i_1(t)=K_1[e^{-pt}e^{j\lambda 1t}]$ (6) , $i_2(t)=K_2[e^{-pt}e^{-j\lambda 2t}]$ (7)

With the boundary conditions and setting $K_1=K_2=I$ is located: $i(t)_{tot}=Ie^{-pt}\ [e^{j\lambda 1t}+e^{-j\lambda 2t}]/2$ (8) Therefore: from [2] pg-48: $i(t)_{tot}=Ioe^{-pt}\ \cos(\omega_o t+\phi)$ (9), with a peak int = 0 and the asymptotic decrescrenza. The pulsation $\omega_o=1/\sqrt{CL}$ .dipende from components CL .

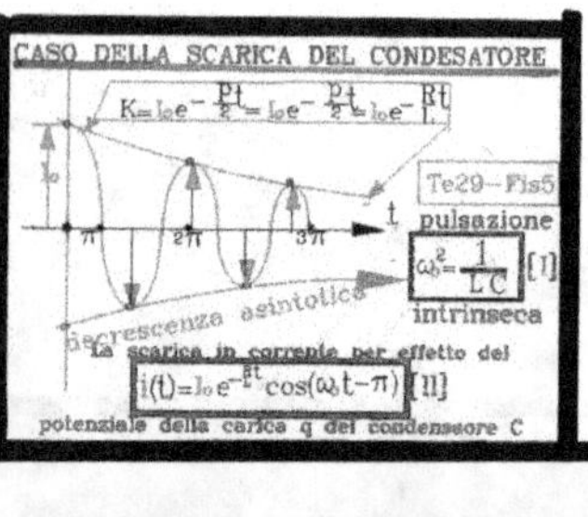

Le proprietà geometriche del raggio d luce coinvolgono un problema che l'Homo sapiens non potrà mai risolvere. Voglimao dire che non potrà mai rispondere alla domanda : A quale velocità si propaga la luce? Questo vale per tutte le grandezze fisicche che oinvolgono energia e materia . Nell'antichià si riteneva che, ad es. la luce si propagasse a velocità infinita. Oggi con l'apporto del ragguio Laser si è determinato il valore c=299.999,54 Kms^{-1} .. Ma è una velocità media o , se si vuole probabile, approssimata allo 0,02 % . Si tenga dunque presente che nel seguito quando si

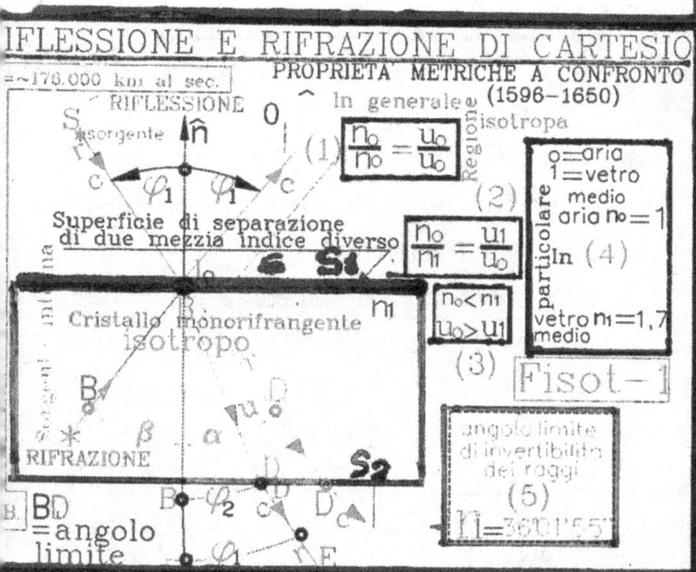

parla di graandezze fisiche come la velocità rappresentata nella Fisot-Fig1 con riferimento alla legge di Cartesio ::

$$n_{12}=\frac{n_1}{n_2}=\frac{\sin\varphi_1}{\sin\varphi_2}=\frac{u_2}{u_1} \quad (1)$$

e di valori metricamente esspressi si tratta sempre di valori approssimati

P1- Analisi della legge di Cartesio

Se dalla sorgente S parte un raggio r luminoso che si propaga nel semiṡpazio C (indice di rifrazione n1, fino alla S di separazione con il crisallo, indice $n_2>n_1$) Allora per il segnale r , giuto in Io sono possibili tre diverse evoluzionii in base allo stato della S,previste dalla (1) .Possono vertificarsi tre casi .

AMEDEO AVOGADRO 1776–1856

Land geometric properties of the beam of light involving a problem that Homo sapiens will never solve. Voglimao say that it will never answer the question: At what speed does light propagates? This applies to all sizes fisicche oinvolgono that energy and matter. Nell'antichià it was believed that, eg. the light will propagate with infinite speed. Today with the help of Laser ragguio has determined the value c = 299,999.54 Kms^{-1}. . But it is an average speed or, if you will likely approximate al- 0.02%. Remember to consider that later when we talk about physical graandezze as the speed shown in Fig1-Fisot with reference to the law of Descartes ::

$$n_{12}=\frac{n_1}{n_2}=\frac{\sin\varphi_1}{\sin\varphi_2}=\frac{u_2}{u_1} \quad (1)$$

and values metrically esspressi it is always approximate values .

P1-analysis of the law of Descartes

If the source S the radius r of light that propagates in the half-space C (refractive index n1-tion, until the S parting with cristal to, index n2> n1)

Then for the signal r, down to I am in three possible evoluzionii based on the state of S, provided (1). may vertificarsi three cases.

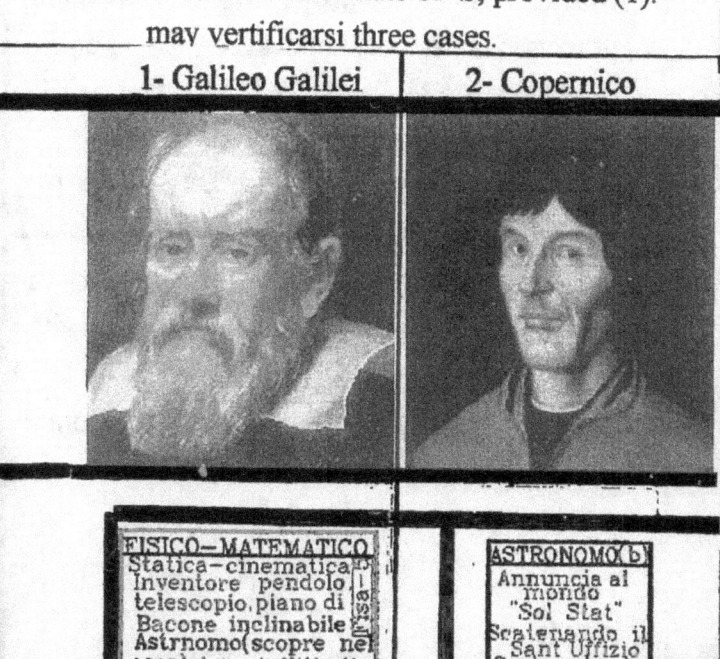

1- Galileo Galilei	2- Copernico
	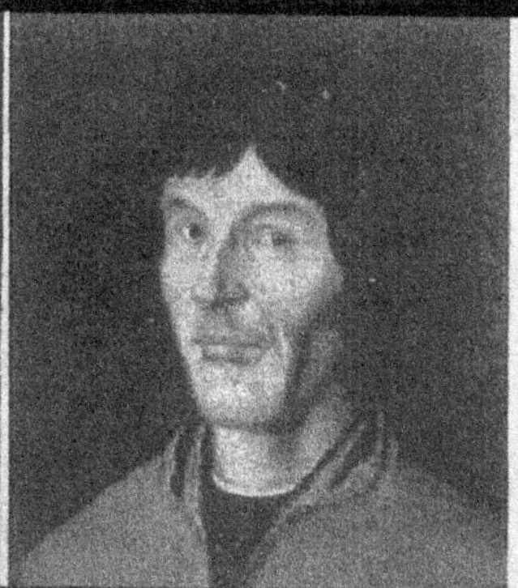

FISICO–MATEMATICO
Statica–cinematica
Inventore pendolo telescopio, piano di Bacone inclinabile Astrnomo(scopre nel 1610) i 4 satelliti di Giove :Io,Europa,Callisto,

ASTRONOMO(b)
Annuncia al mondo "Sol Stat" Scatenando il Sant Uffizio Sostenuto da Galilei che solo per questo

Riprendiamo la **Fig1-Fisot1** , supporto al teorema di **Cartesio** della eq.((1)),ora dotata del rapporto polifrequenziale dei segnali sinusoidali :

$$n_{12}=\frac{n_1}{n_2}=\frac{\sin\varphi_1}{\sin\varphi_2}=\frac{u_2}{u_1}=\frac{f_2}{f_1} \quad (1)$$

Le radiazioni **f** provenienti dal **Sole** sono di luce bianca polifrequenziale.

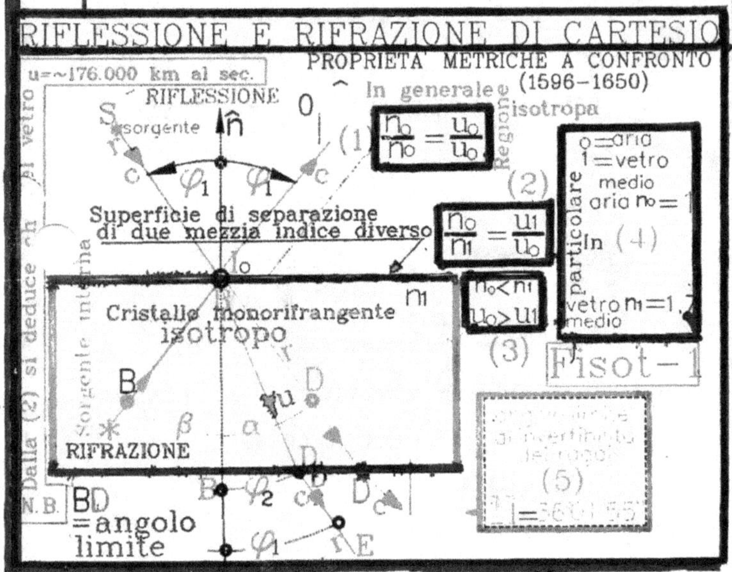

Ad es. il **rosso** di ($3,75 \cdot 10^{14}$ **Hz**) e il **violetto** di ($7,5 \cdot 10^{14}$ **Hz**). Valori riferiti alla propagazione nel vuoto e sensibilmente nell'aria

#P2- Analisi della legge (1) generalizzata

1-Il raggio **r** parte dalla **sorgente S** segue il percorso **S→ I₀→ D→ E** (2) . In tal caso **r** è rifratto poichè nel tratto **S→ I₀** ≡ **r** la traccia rettilinea si propaga nella materia con indice di rifrazione n_1 . Nel tratto **I₀ → D ≡ r** il raggio attraversa il cristallo supposto con indice n_2 Infine nel tratto **D→ E≡ r ≡ S→ I₀** indice n1.

2- Nel percorso (2) **r è rifratto in Io e D**

$$n_{12}=\frac{n_1}{n_2}=\frac{\sin\varphi_1}{\sin\varphi_2}=\frac{u_2}{u_1}=\frac{f_2}{f_1} \leftrightarrow n_{21}\frac{n_2}{n_1}=\frac{\sin\varphi_2}{\sin\varphi_1}=\frac{u_1}{u_2}=\frac{f_1}{f_2} \quad (3)$$

Lo spettro ottico delle frequenze e dei colori di luce bianca emessa dalla **sorgente S** è in realtà a carattere distribuzionale evidenziato dalle gocce d'acqua(microcristalli)dopo un temporale

Resume the **Fig1-Fisot1**, support theorems of **Descartes** but of eq. ((1)), now equipped with the ratio of polifrequenziale segnalisinusoidali:

$$n_{12}=\frac{n_1}{n_2}=\frac{\sin\varphi_1}{\sin\varphi_2}=\frac{u_2}{u_1}=\frac{f_2}{f_1} \quad (1)$$

The radiation coming from the Sun are of white light polifrequenzial.

Eg. the **red** ($3,75 \cdot 10^{14}$ Hz) and **violet** ($7,5 \cdot 10^{14}$ **Hz**). Amounts refer to the propagation in vacuum and in the air significantly.

 # P2-Analysis of the Lawgeneralized (1)

1-The radius **r** part follows the path from the **source S** side **S→ I₀→ D→ E** (2).

In this case **r** is refracted as in **S→ I₀ ≡ r** the straight track propagates in the material with refractive index n_1.

In section **I₀ → D ≡ r** assumed the beam through the crystal with index n^2

Finally, in section **D→ E≡ r ≡ S→ I₀** index n1.

2 - On the way (2) **r is tefracted in Io and D**

$$n_{12}=\frac{n_1}{n_2}=\frac{\sin\varphi_1}{\sin\varphi_2}=\frac{u_2}{u_1}=\frac{f_2}{f_1} \leftrightarrow n_{21}\frac{n_2}{n_1}=\frac{\sin\varphi_2}{\sin\varphi_1}=\frac{u_1}{u_2}=\frac{f_1}{f_2} \quad (3)$$

The optical spectrum of frequencies and colors of white light emitted by the **source S** is actually a distributional character evidenced by drops of water (micro cristal) after a storm.

R. CARTESIO

1596-1650

Isaac Newton è stato uno dei primi ricercatori ad occuparsi dei fenomeni luminosi Nel 1666 infierendo la peste a Cambridge, dove insegnava , si è rifugiato nella fattoria materna. Questo gli permise di fare un esperimento su un problema che lo affascinava, La luce è quel fenomeno misterioso dal comportamento strano

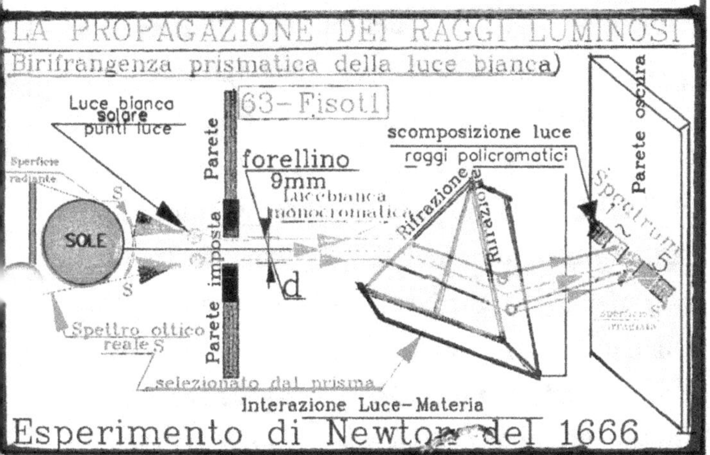

Fatto un foro di ~ **9 mm** sulla imposta della camera. Chiuse porte e finestre . Vide sulla parete opposta proiettarsi un pacchetto di raggi di luce bianca. Interposto un prisma di Spato d'Islanda. Ha visto disegnarsi nella parete opposta la luce filtrata dai prisma (unendo due prismi con balsamo del Canada) Fig19-Stru1. vide (miracolo) com- parire una distribuzione di colori . Ma levato il prisma la luce a colori sparisce. Perciò dette al fenomeno il nome di Spectrum . Per Newton il problema nasce per stabilire la natura di quei "raggi solari" Ricordiamo che Newton nel 1661 venne iscritto al Trinity College di Cambridge , alla età di anni19. Scolaro alquanto più anziano degli altri e meno colto ma molto maturo Infatti nel 1685 presenta ai soci della rivista Royal Society il suo capolavoro : **Philosophiae naturalis principia mathematica.** La sintesi è formalizzata in $\vec{F} = g \frac{m_1 m_2}{r^2} \underline{r}$ (1) fra due masse pesanti m_1 e m_2, g=gravità r = versore unitario o direzione.

HUYGENS

1629-1695

Isaac Newton was one of the first researchers to deal with the luminous phenomena.

In 1666 to treat la plague in Cambridge, where hegoes, took refuge in the farm nursery. This allowed him to make an experiment on a problem that fascinated him. ,

The light is the mysterious phenomenon by the strange behavior

Made a hole of ~ 9 mm on the sets of the room. Chiuse doors and windows. He saw on the opposite wall to project a package of white light rays. Interposed a prism of Iceland spar. He saw the draw itself in the opposite wall light filtered through the prism (combining two prisms with Canada balsam) Fig19-Stru1. saw (miracle) com-pear a distribution of colors. But it raised the prism light color disappears. Therefore, the phenomenon called the name of Spectrum.

For **Newton**, the problem is to determine the nature of those sun rays © [a] Recall that Newton was enrolled in 1661 at Trinity College, Cambridge, at the age of 19 years. Schoolboy somewhat older than the others and less educated but very mature .

fact in 1685 presents to members of the Royal Society magazine his masterpiece;

: ,

Philosophiae naturalis principia Mathematica.

The synthesis is formalized in

$\vec{F} = g \frac{m_1 m_2}{r^2} \underline{r}$ (1) , with r barycentric distance between two heavy masses m_1 and m_2, g = gravity $\underline{r}$ = vector united o direction

I. NEWTON

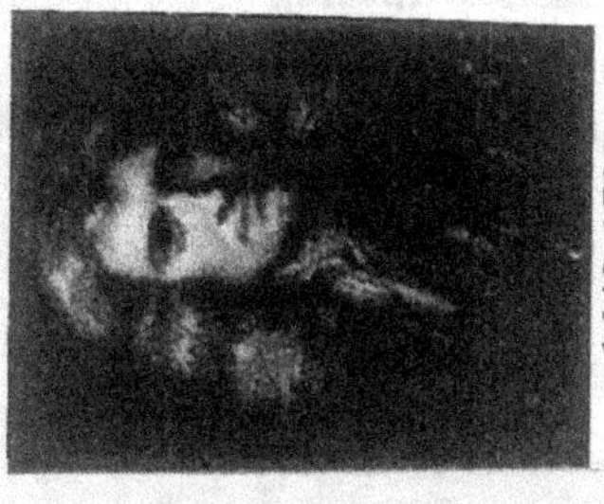

1642-1727

La legge di attrazione universale è nata dopo un percorso lungo e pieno di ostacoli . Infatti fra l'esperimento dello Spectrum(1666) al (1685) in cui annuncia con la pubblicazione della meccaanica celeste la attrazione iniversale sono intercorsi 19 anni. Come spiegare un si lungo intervallo dato dell'annuncio del 1685 ? In primo luogo la meticolosa descrizione dei movimenti delle masse come densità di materia Fedele al principio sperimentale iniziò prendendo a modello per i calcoli la Terra e la Luna impostando la relazione $f=k\frac{Mm}{r^2}$ (1) Conosceva il raggio della Terra calcolato da **Eratostene** circa 2000 anni prima. circa 7200 km e la distanza della **Luna** dalla **Terra** di **Ipparco** circa 420.000 km. Sapeva da **Keplero** che il moto della Terra intorno al sole non era circolare come aveva supposto Copernico ma eccentrico con **eccentricità e=0,17 a** essendo a il semiasse maggiore della ellisse dell'orbita terrestre rispetto al Sole

. Il ragionamento doveva essere presso a poco il seguente. Se considero la luna in posizione sinodica(allineata con la Terra e il Sole) in posizione perigea usando la (1) si può scrivere: $f_1=k\frac{Mm}{r_1^2}$ (2) .In posizione apogea risulta $f_2=k\frac{Mm}{r_2^2}$ (3) dato che le distanze sono tali che **r1(perigea)<r2 (apogea)**

Ignoriamo la dimostrazione di **Newton** ma è noto che interpretò con felice intuizione che la forza f_1 è maggiore della forza f_2 per le masse terrestri mobili rispetto alla terraferma. In conclusione la (1) è fallita per il calcolo della massa M dato che f1 ed f2 dovevano corrisspondere alta e bassa marea . La costante k era nei due casi diversa quindi la (1) era fallita. Non gli rimase altro che deporre nel cassetto il laborioso lavoro La Fig1-Sto1 rende ragione del fatto

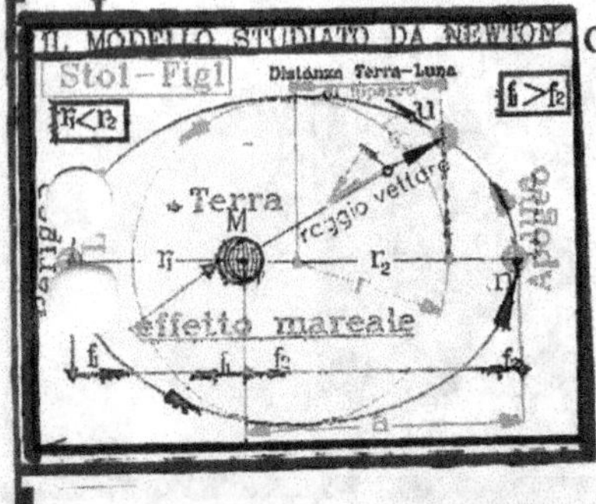

The universal law of attraction is born after a long and full of obstacles. In fact, between the experiment **Spectrum** (1666) to (1685) in which announces the publication of meccaanica celestial attraction iniversale have lapsed 19 years old. How to explain it a long interval since the announcement of 1685? First the meticulous description of the movements of the masses as the density of matter Faithful to the principle sperimentale began a model for the calculations the Earth and the Moon setting up the equation $f=k\frac{Mm}{r^2}$ (1).. He knew the radius of the Earth calculated by **Eratosthenes** about 2000 years ago. about 7200 km and the distance of the **Moon** from the **Earth Hipparchus** about 420,000 km. .To know by Kepler movie that the Earth around the sun was not circulate as **Copernicus** had assumed but eccentric with **eccentricity: e = 0.17a** being the semi-major axis of the ellipse terrestrial orbit relative to the Sun's reasoning had to be very nearly as follows. If I consider the moon in synodical position (aligned with the Earth and the Sun) in position perigea using (1) we can write: $f1=k\frac{Mm}{r_1^2}$ (2).

Ideally apogea is $f2=k\frac{Mm}{r_2^2}$ (3),

since the distances are by nature such That:

(perigea) r1 < (apogea) r2

ignore the demonstration of Newton but it is known that he played with intuition that the force is greater than the force f_1 f_2 for the masses land mobile than the mainland.

In the conclusion (1) failed for the calculation of the masses M since f_2 as low tide. But the constant k in the two cases was different then the (1) had failed. He had no choice other than laying in the drawer the laborious work The Fig1-STO1 accounts for the fact that if the lunar orbit was circular (1) would be exact. But

IPPARCO

190-120 a.c.

I. NEWTON

1642-1727

ERATOSTENE

275-195 a.c.

La legge di attrazione universale enunciata nel 1685 era stata formalizzata come principio anni prima e che conosciamo: $f = k\frac{Mm}{r^2}$ **(1)**.

Abbandonata per il mancato riscontro sperimentale. Cherchiamo ora con il supporto della Fig-1 il perchè della causa.

P1-Analisi della (1) Le masse della Terra e della Luna sono i dati del problema di fronte ai quali si trovò **Newton.** Per la determinazione della M della Terra. Trovata la densità madia ρ della M.Mancava allora il volume dello sferoide terrestre. Conoscendo la determinazione di **Eratostene** del raggio R terrestre (~ R=7.000 km).Allora M: $M = \rho V = r((4/3)\pi R^2$ (2) essendo ρ densità per km^3 della Terra.

Ma inserita la (2) ,cioè M nella **(1)** ne riscontrò il fallimento dell'effetto mareale lunare **m** sulla **M**. Da qui il fallimento. Nel 1670 **Jean Picard** astronomo e geodeta (1620-1682)calcolò un arco di meridiano fra Sourdon e Malvoisine da cui ricavo il raggio R della Terra supposta sferica più preciso di quello di **Eratostene** di 19 secoli prima. Era quello che **Newton** attendeva Ripresa la (1) verificò con (2) il sincronismo mareale **Terra-Luna** .

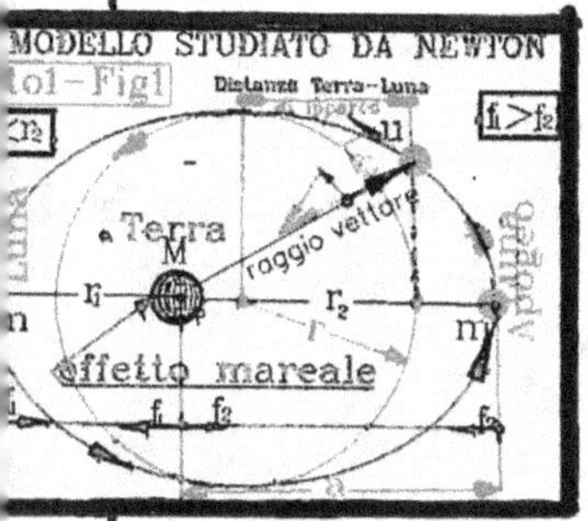

Ignoriamo lo sviluppo matematico di Newton Sostanzialmente si procede nel seguente modo:

-La **Luna** sia al **perigeo**, allora: $f_1 = k\frac{Mm}{r_1^2}$ **(3)**

-La **Luna** si trovi all'**apogeno** : $f_2 = k\frac{Mm}{r_2^2}$ **(4)**

Dato che $r_1 < r_2$ (5) il nostro satellite eserciterà $f1 > f2$ (6) . La max attrazione della m sui mari M(precisamente sulle acque rispetto alla terraferma), ha per causa la (3), quando la **Luna** si trova al **Perigeo** con f_1 max e f_2 con la **Luna** all'**Apoge** a causa della **(4)** ,giusto per le distanze **(5)** ,in sincronia con l'effeto mareale

The law of attraction universal enunciated in 1685 was formalized as a principle and years before we know: f = $f = k\frac{Mm}{r^2}$ **(1)**. Abandoned for failure to reply experimental. Io considerer now with the support of the Fig-1 because of the cause

P1-Analysis (1) The masses of the Earth and the Moon are the data of the problem to which he found himself in front of **Newton**. For the determination of M of the Earth. Found the density ρ of the cupboard M.Mancava then the volume of the terrestrial spheroid. Knowing the determination of the radius R of **Eratosthenes** terrestrial (~ R = 7000 km). Then M:

$M = V\rho = \rho((4/3)\pi R^2)$ (2) being for km^3 ρ density of the Earth.

But inserted (2), wich M in **(1)** will be set against the failure of the effect on tidal lunar **m** in **M**.

Hence the failure. In 1670, surveyor and astronomer **Jean Picard** (1620-1682) calculated an arc of the meridian between Sourdon and Malvoisine from which proceeds the radius R of the spherical Earth suppository more precise than **Eratosthenes** of 19 century before. It was what I expected Returning to the **Newton** (1) occurred with (2) the Earth-Moon tidal synchronism.

We ignore Newton's mathematical development is essentially proceeds as follows

-The **Moon** is at **perigee**, then: $f_1 = k\frac{Mm}{r_1^2}$ **(3)**

-The **Moon** is at **apogee** then $f_2 = k\frac{Mm}{r_2^2}$ **(4)**

Since $r_1 < r_2$ **(5)** the our satellite will exercise $f1 > f2$ (6) The attraction of max m in the sea of M (specifically on the water than on land), has for the cause **(3)**, when the **Moon** is at **perigee** and **apogee** wich f_1 and f_2 max and is due to **(4)** for the distances (5) in synchrony with the tidal effects are normally on Earth

Riprendiamo quanto lasciato in sospeso nella pagina 16 con lo **Spectrum** , Fig63 Fisot1 Dopo le conclusioni delle interazioni fra mate-ria e materia dei corpi pesanti di cui alla legge omnicomprensiva : $F = k\frac{Mm}{r^2}$ (1) , estesa a tutte le masse pesanti,dal-le polveri sottili ai quasar (ammassi stellari agli estremi dell'universo) nel problema dei raggi luminosi la interazione è fra energia e materia. Nel formulare la ipotesi sulla natura della luce **Newton** senza perplessità intepreta la luce come delle particelle materiali irradiate dalle stelle o da corpi luminosi come i metalli portati alla incandescenza. E fino a questo punto la sua visione ottica del problema non si discosta dalla legge di attrazione (1) La prima obbiezione è che ad opera del cristallo queste particelle dovrebbero serapare la loro micromassa Manca ll riscontro e la interpreta-zione risulta lacunosa. **C.Huygens**(1619- 1695) in senso opposto espone la teoria ondulatoria della luce. Si tratta dunque di un fenomeno periodico che si propaga nello spazio .Privo di massa materiale ,quindi di fenomeno di pura energia , Fig54-Fis4 di onde elettromagnetiche {E ,H} . Come si può costatare i due modelli a confronto rivelano la fondamentale diversità Il modello nwtoniano è radiale ed interagisce su supporto euclideo fra M ed m.Diversamente per **Huygens** il supporto è sferico e legato al tempo reale quadridimensionale. Si diffonde per onde di vario ordine (periodiche)

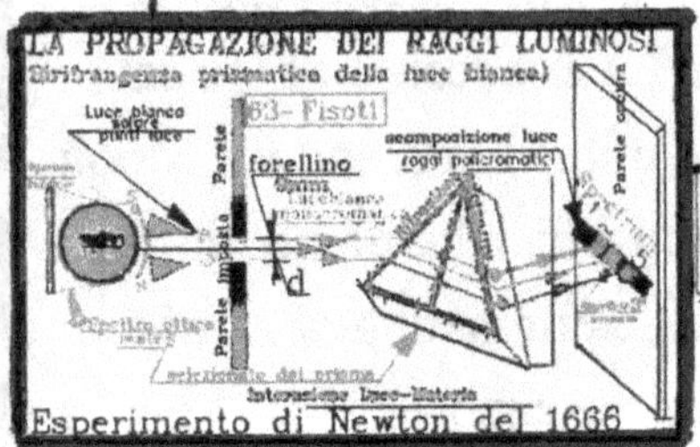

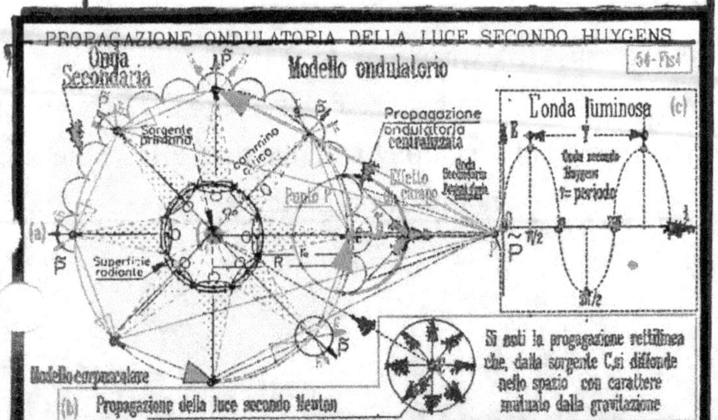

We take the left off on page 16 with the **Spectrum**, Fig63 Fisot1 After the conclusion of the interactions between matter and mate-ria of heavy bodies referred to in Law encompassing:

$F = k\frac{Mm}{r^2}$ (1), extended to all heavy masses, from the fine-quasars (star clusters at the ends of the universe) in the problem of the light rays is the interaction between energy and matter. In formulating hypotheses about the nature of light **Newton** without perplexity interprets light as material particles radiated by the stars or luminous bodies such as metals brought to incande-scence. And up to this point its optical vision of the problem does not deviate from the law of attraction (1) The first objection is that the crystal by these particles would serapare their micromassa missing'll reply and the interpretation is lacunosa. **C. Huygens** (1619 -1695) in the opposite direction exposes the wave theory of light .

Si is therefore a periodic phenomenon which propagates in space.

Devoid of material mass, then the phenomenon of pure energy, Fig54-Fis4 of electromagnetic waves{E ,H}

As can be seen comparing the two models reveal the fundamental difference nwtoniano The model is radial and interacts support of Euclidean between M and m.Diversamente for **Huygens** support is spherical and linked to real-time four-dimensional. It radiates waves of various kinds (periodic)

HUYGENS

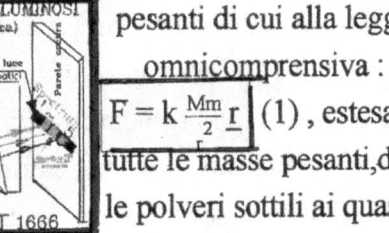

1629-1695

La ipotesi ondulatoria in contrato con la corpuscolare deve aver scoperto i nervi di **Newton** . Infatti si esprime in proposito con questo giudizio

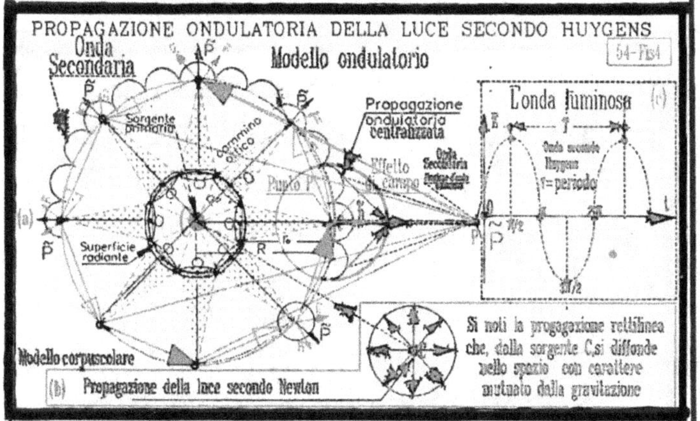

a dir poco spregiativo << Dire che la luce è un fenomeno ondulatorio come dicono certi filosofi è...... >> . Non è la sola ombra. Ricordiamo che negli anni di insegnamento di fisica matematica in Cambridge **Newton** riusci a sviluppare una serie geometrica convergente a 2, caso particolare :

$$1+\frac{1}{2}+\frac{1}{2^2}+....+\frac{1}{2^n}=\lim_{n\to\infty}\frac{1-0,5^n}{1-0,5}=2 \quad (1)$$

. Questa serie è stata la causa di una accusa di plagio diretta all'amico **G.W.Leibnitz**(1646-1716) , insigne matematco.Fondatore del calcolo integrale.In particolare . Integrando la funzione f(x)=x, fra 0 e 2 Con **Leibnitz** si ottiene : $\int_0^2 xdx=\frac{x^2}{2:}=4:2=2$ (2) .

Newton per interposta persona accusa **Leibnitz**. di aver plagiato la (1) . Non è difficile capire che la differenza fra la (2) ,seppur formalmente di valore uguale alla (1) e un'altra cosa Infatti. La(1)è una serie numerica di una successione di infiniti addendi .Inoltre la progenitura spetta a **A. Taquet**(1612-1660), della compagnia di Gesù che nel Suo **Cilindricorum,annularium libri quator** del 1656 si trova per la prima volta , in un libro , **la serie:** $\sum_0^\infty aq^n=\frac{a}{1-q}$ (3) Se q<1 , e la a ≠ 0

La Figura 34-Fis1 mostra (a) di **Newton** → (1)

La (2) in generale è del tipo: $\int f'(x)dx=F(x)+C$ (4) ,con f'(x) primitiva di F(x) e C una costante arbitrariai integrazione. La (4) esclude che ci sia plagio!

The wave hypothesis in contract with the corpuscular must have discovered the nerves of **Newton**. It is expressed in the way with this judgment to say the least derogatory

<< To say that light is a wave phenomenon as some philosophers say is ...... >>.

Is not the only shadow. Recall that in years of teaching mathematical physics at Cambridge **Newton** was able to develop a convergent geometric series at 2, special

$$1+\frac{1}{2}+\frac{1}{2^2}+....+\frac{1}{2^n}=\lim_{n\to\infty}\frac{1-0,5^n}{1-0,5}=2 \quad (1)$$

This series is the cause of a direct accusation of plagiarism friend **G.W. Leibnitz (1646-1716)**, a distinguished matematco.Fondator the calculation integrale.

In particular. Integrating the function f (x) = x, between 0 and 2 . With Leibnitz is obtained:

$$\int_0^2 xdx=\frac{x^2}{2:}=4:2=2 \quad (2)$$

. (2).

Newton nominees accused **Leibnitz** having plagiarized (1). It is not difficult to understand that the difference between (2), although formally a value equal to (1) and another thing fact. Equation (1) is a numerical series of successive infinite summands.

Moreover, the offspring belongs to **A. Taquet** (1612-1660), the company of Jesus and His **Cilindricorum et annularium Quator books** of 1656 is for the first time in a printed book, the series:

$$\sum_0^\infty aq^n=\frac{a}{1-q} \quad (3) \quad \text{Se } q<1 , a\neq 0$$

In Figure 34 -Fis1 shows (a) of **Newton** → (1)

The (2) in general is of the type: $\int f'(x)dx=F(x)+C$ (4) with f '(x) primitive F (x) and C an arbitrary constant of integration.

The (4) mean that there is plagiarism!

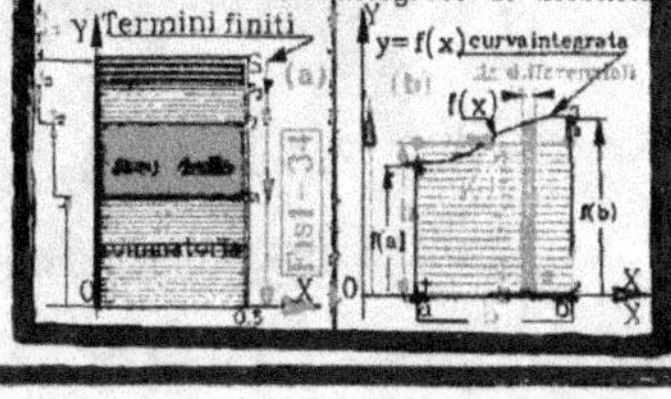

La grandezza di **Newton** la Old England lo pianse come un figlio.Tumulato in Westmister.Il feretro, portatoda 4 pari. Amicie ed ammitatotori gli hanno dedicato questo epitaffio: **H.S.E** (a)

ISAACUS NEWTON, EQUES AURATUS QUI ANIMI VI PROPRE DIVINA PLANETARUS MOTUS,FIGURAS, COMENTARUM, SEMITAȘ OCEANIQUE AEST,SUA MATHESIS FACEM PRAEFERENTE,PRIMA DEMOSTRAVIT. RADIORUM LUCIS DISSIMILITUDINES,COLORUM INDE NASCENTIUM PROPRIETATES QUAS NEMO ANTEA VEL SUSPICATUS ERAT,PREVEVETIUCAVIT NATURE,SEDULUS ,SAGAX,FIDUS INTERPREPRES, DEI OPT MAX MAJESTATEM PHILOSOPHIA ASSERUIT,EVANGELEII SEMPLICITTEM MORIBUS ESPRESSIT . SIBIATULENTUR MORTALES,TALE TANTUNQUEEXTETISSE HUMANI DECUS GENERIS.

NATUS xxv DECEMB MDCXLII OBIT MAR.MDCCXXVII

H.S.E (b)

ISACCO NEWTON CAVALIERE D'ORO CON FORZA D'ANIMO ,QUASI DIVINA, DIMOSTRO' IL MOVIMENTO DEI PIANETI CON DISEGNI E NOTE PERTINENTI(commentarium) DESCRISSE LE ROTTE NAVALI NEGLI OCEANI E LA CONFIGURAZIONE MATEMATICA) lA DIFFERENTE LUCE DEI RAGGI DEL SOLE CHE NESSUNO MAI PRIMA AVEVA INTUIT0. SOLLECITO, SAGACE ,INGEGNO DICHIARO'

CON LS FILOSOGIA LA MAETA' DI DIO. OTTIMO MASSIMO.ASSERI' LA SEMPLICITA' DEL VANGELO E NEI COSTUMI(PORTO') TANTA DIGNITA' AL GENERE UMANO MAI VISTA PRIMA

Gli amici giustamete ricordano che **Newton** nel 1666,infierendo la peste in Cambridge, si è rifugiato nella fattoria materna realizzando un esperimento covato da tempo.Fatto un buco nella imposta,Fig65. Interposto un prisma cristallino scopre la **diffrazione dei raggi luminosi** e disppiegarsi sulla parte oscura un arcobaleno di colori. E' la prima volta nella storia che un uomo riesce a riprodurre in laboratorio di che la natura mostra dopo i temporali sul cielo dispiegarsi dei colori. **Newton** lo battezza con il nome evocativo di **Spectrum** perchè sparisce al levar del **prisma**

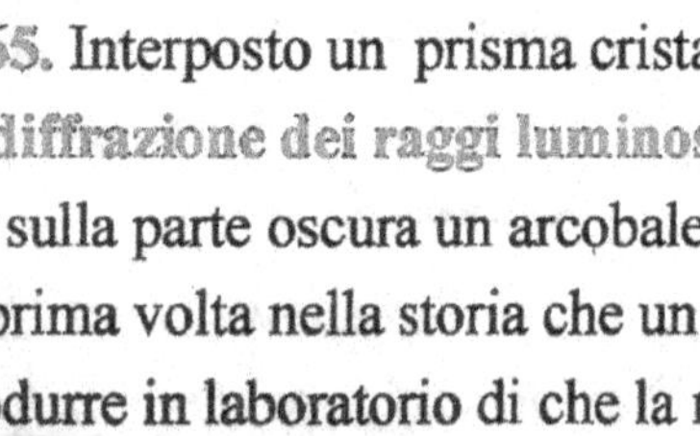

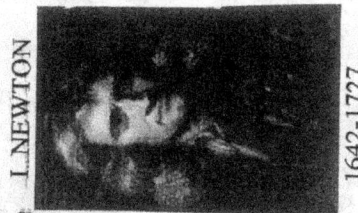

The magnitude of the **Newton** Old England the floor-as if in a figlio.Tumulato Westmister.Il fere-tro, portatoda 4 peers. Amicie and ammitatotori have dedicated this epitaph: HSE (b)

SAAC NEWTON GOLDEN KNIGHT with fortitude, almost divine, SHOWS 'THE MOVEMENT OF THE PLANETS WITH DRAWINGS AND RELEVANT NOTES (Commentarium) described the SHIP ROUTES IN THE OCEANS AND THE CONFIGURATION OF MATHEMATICS) lA DIFFERENT LIGHT RAYS OF THE SUN THAT NOBODY EVER FIRST HAD INTUIT0. REMINDER, SAGACE, INGEGNO DECLARE '

WITH THE LS FILOSOGIA MAETA 'OF GOD. EXCELLENT MASSIMO.ASSERI 'SIMPLICITY' THE GOSPEL AND THE COSTUMES (PORT ') SO MUCH DIGNITY' THE HUMAN RACE NEVER SEEN BEFORE

Friends giustamete recall that **Newton**, in 1666, the plague raging in Cambridge, it was refuge the farm maternal performing an experiment hatched by tempo.

Fatto a hole in the tax Fig65.

Interposed a prism lens discovers the **diffraction of light rays** on the dark side and disppiegarsi a rainbow of colors.

It 's the first time in history that a man is able to **reproduce in the laboratory** of nature shows that after storms on the sky unfolding of colors.

Newton baptizes him with the evocative name of **Spectrum** because it disappears at the rising of the **prism**

La legge di attrazionen di Newton ,rappresentata Nella Fig-5 e formalizzata :

$$f_1=k\frac{Mm}{r_1^2} \quad (1)$$

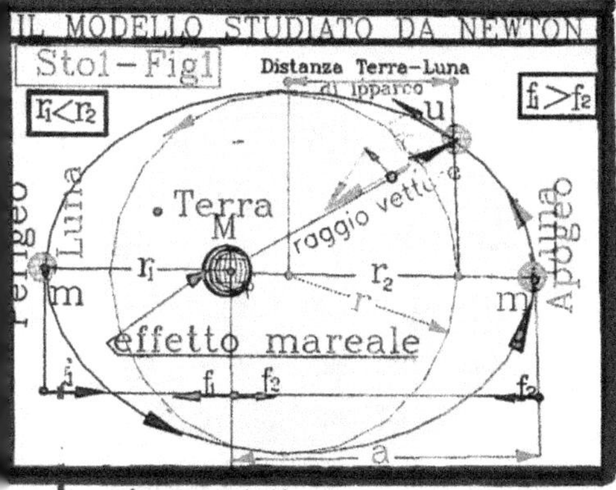

associata alla :

$$f_2=k\frac{Mm}{r_2^2} \quad (2)$$

di fatto chiude un periodo storico nel quale la ricerca della conoscenza del sapare per il sapere è condizionato dal dominio temporale religioso. Nella cronologia storica abbiamo suddiviso la " Storiografia Scientifica" in tre periodi distinti legati a situazioni ambientali dalle caratteristiche diverse e dalla tecnologia associata in continua evoluzione Il periodo Indo-Babilonese di incerta datazione ma sostanzialmente, sulla base dei referti archeologici e bibblici dal 2000 a.C.al 200 d.C.In questo periodo arcaico la ricerca scietifi-ca era a carattere astronomico e speculativo. Ciò è stato dimostrato da una tavoletta di terracotta con la stella Sirio(che guidò i Re Magi verso Bethlemme .Luminosità ~m=+0,2 contro m= -17 del Sole , PogsonAsr1) rinvenuta in Mesopota- ia (attuale Bagdad) per poi diffondere in Grecia , Egitto,Italia meridionale. Con la caduta di Sira- cusa ad opera del Proconsole romano Marcello (212 a.C.) si Chiude Con Archimede l'evo antico. Infatti Archimede ,inventore del primo cannone detto Catapulta , è stato ucciso (gli storici affer- affermano che un solato romano lo uccise, non avendolo riconosciuto),Versione poco credibile per e il fatto che Archimede in difesa di Siracuss scagliava sassi e olio bollente contro gli assalit0-ri Con ciò spegnedo la luce scientifica della anti-ca Grecia . Il S.P.QR domina su tre continenti

The law of attrazionen Newton, represented in Fig-5 and formalized:

$$f_1=k\frac{Mm}{r_1^2} \quad (1)$$

associated with:

$$f_2=k\frac{Mm}{r_2^2} \quad (2)$$

In fact [illegible] historical period in which the pursuit of knowledge of Useful facts for knowledge is conditioned by time domain religion. In the historical chronology we have divided the Scientific [a] Historiography in three distinct periods related to environmental situations with different characteristics and the associated technology evolving Indo-Babylonian period of uncertain date but basically, on the basis of archaeological reports and bibblici since 2000 a.C and 200 d.C in the archaic period this research was to scietifi-ca astronomical character and speculative. This has been demonstrated by a terracotta tablet with the star Sirius (which guided the Magi to Bethlehem. Brightness ~ m = 0.2 m = -17 against the Sun, Pogson Astr1) ia found in Mesopotamia (now Baghdad) to then spread to Greece, Egypt, southern Italy. With the fall of Sira-accusation by the Roman proconsul Marcello (212 BC) Closes With Archimedes the ancient ages. In fact, Archimedes, inventor of the first Catapult Cannon said, was killed (the historians affirm affirm that a Roman consulate killed him, not having recognized), Version implausible given the fact that Archimedes in defense of Siracu-sa hurled stones and hot oil against the attackers. With this spegnedo the scientific light of ancient Greece. The SPQR dominates three continents.

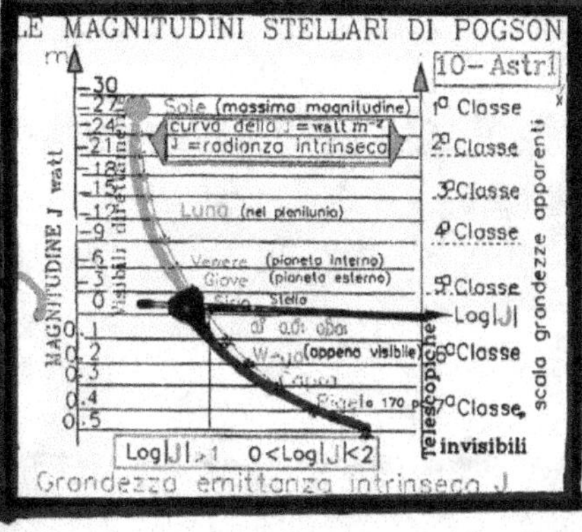

POGSON e ls MAGNITUDINE

ROMA URBI ET ORBI(I mala tempora) pg-23

Chi è quel personaggio che va predicando la eguaglianza fra gli uomini ? Questa domanda è posta al S.P.Q.R.della Roma Caput mundi. La filosofia di **Cristo** è un pericolo per Roma e ordina alle sue falangi dislocate in Terra di Israel di imporre al Sinedrio di crocifiggere **Cristo.** Il Sinedrio è un potere disarmato ed è costretto ad ordinare la crocefissione di **Cristo** fra i ladroni perchè serva di monito.La storia di Cristo uomo nella fede ha lasciato una impronta più profonda e duratura di ogni altro nei popoli cristiani . Ai tempi di **Roma Caput mundi** era pericoloso manifestare sentimenti cristiani ma ciò servi a ingrossare le file degli apostoli e dei martiri. Lo sfacelo di Roma avviene con il concorso della potenza posseduta (una belva che divora se stessa creando ricchezza e corruzione. corruzione) con il concorso della fede che ha finito per dare nuova forma alla romanità cristiana Roma in breve diventa il centro urbi ed orbi. Ma in quanto umana è soggetta ,dopo la nascita e la crescita della potenza alla decadenza della fede. Qui prendiamo l'avvio,collegandoci ai personaggi del quadro iconografico di pg-28 (sintesi iconografica storiografica) Notiamo che il Solo **Newton** non ha subito violenza dalla inquisizione. Diversamente **R.Bacone** e **Galilei** sono stati incarcerati. Si deduce che il potere spirituale può mutare la fede in fanatismo. La prova è codificata nella sentenza del Santo Ufficio che manda sul rogo per la purificazione dei peccati il fisolofo **G.Bruno** da Nola.Arrestato dal Mocenigo a Venezia per il sospetto di stregoneria.Consegnato a **Roma Urbi ed Orbi**. Concluse la sua vicenda terrena affrontando la morte con dignità . Per **G.Brunoo le religioni sono forme di potere spirituale volte ad oscurare la realtà di Dio , che si identifica con la natura!Era una notte dell'anno Domine 1600**

Di quella Pira l'orrendo fuoco!

ROME URBI ET ORBI (I mala tempora) pg-23

Who is that person who has been preaching the equality of men? This question is posed to S.P.Q.R.of the Roma Caput Mundi. The philosophy of **Christ** is a danger to Rome and ordered his phalanges located in the Land of Israel to impose the Sanhedrin to crucify **Christ**. The Sanhedrin is a power unarmed and was forced to order the crucifixion of **Christ** to the thief-ing it out as monito.La history of man in the faith of **Christ** has left a deep and lasting impression more than any other in the Christian peoples. At the time of **Rome Caput Mundi** was dangerous to express Christian sentiments but what servants to swell the ranks of the apostles and martyrs. The collapse of Rome is done with the course-with the power held (a beast that devours itself by creating wealth and corruption. Corruption) with the help of faith that has fi-nite to give new shape to the Roman Christian Rome soon became the center urbi and orbi. But as a human subject, after the birth and growth of power to the decline of faith. Here we take the start, connecting us to charac-gi's iconographic picture of pg-28 (iconographic synthesis historiographical) We note that the only **Newton** has not been subjected to violence by inquisition. Unlike **R.Bacone and Galileo** were imprisoned. We deduce that the spiritual power can change the faith in fanaticism. The proof is encoded in the judgment of the Holy Office, which sends the stake for the purification of sins by the fisolofo G.Bruno Nola.Arrest from Mocenigo in Venice on suspicion of stregoneria.Delivered **Urbi and Orbi** in Rome. He ended his earthly facing death with dignity.

For **G.Brunoo religions are forms of spiritual power-rare times to obscure the reality of God, which is identified with nature was a night of the year 1600 Domine**

Of that Pira the horrible fire!

GIORDANO BRUNO

1548-1600

Alla caduta di Roma Caput Mundi e la meteora Araba,il mondo occidentale è dominato da Roma Urbi et Orbi. Nonostante l'ombra del potere temporale della chiesa di Roma ,inizia un prodiguiso sviluppo scientifico.I poli scientifici di Babilonia Alessandria d'Egitto ,Atene sono sostituiti dai poli diToledo.Firenze, Wejl , Praga, Cambridge Questo dominato dal fondatore della meccanica celeste ,I.Newton (1643-1727)

LA ASTRONOMIA DEI PADRI FONDATORI

Grava l'ombra inquietante del Santo Ufficio. Nonostante questo è in particolare la astronomia di posizione che spicca il volo verso il cielo. ed è fatale il cozzo con la spiritualità della fede nella religione dogmatica. L'uomo vuole capire dove si trova nell'immensità del creato, incredulo del dogma che pone la Terra al centro dell'Universo. Ecco l'uomo del destino, a 19 secoli da Ipparco.

Uno sonosciuto,Tale **N.Copernico** (1473-1543), nato a Thorm (Polonia), Studiò a Cracovia e poi alla università di Bologna(1496) seguendo i corsi di teologia ed astronomia.

Nel 1501è nominaio canonico in Frauenburg.Si ferma in Italia. A Padova dove studia medicina e Ferrara diritto canonico. Dal 1504 si stabilisce definitivamente in Emerlamd (Prussia orientale) dove, in prevalenza, si occupa di astronomia..Nel 1514 interpellato dal Concilio Laternense sulle fasi lunari , non essendo certo della posizione degli astri Luna-Sole-Terra tace (Sto2-Fig4) Dopo numerose osservazioni con base Terra con Marte rileva che è il Sole fermo. Ma per timore di Roma Caput **Urbi et Orbi** non da l'annuncio " SOL STAT"

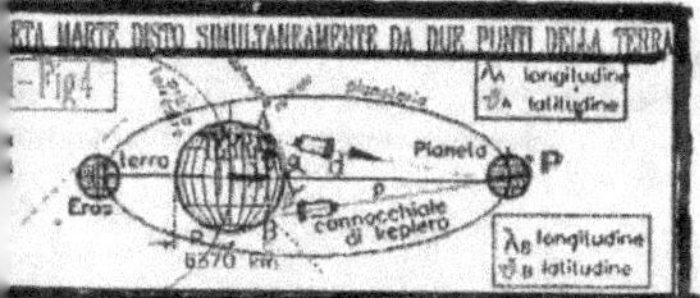

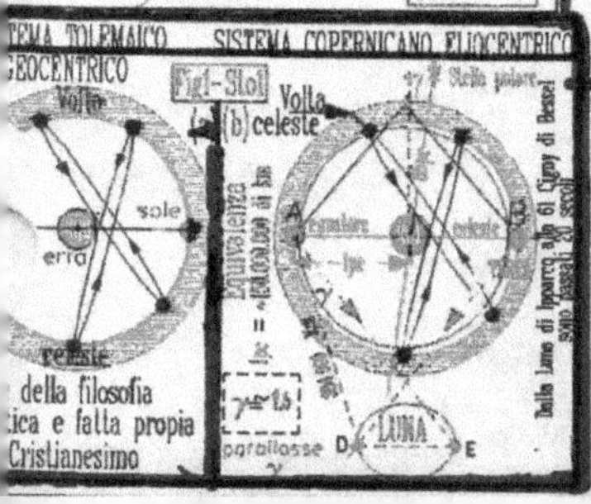

After the fall of Rome Caput Mundi and the meteor Arab, the Western world is dominated by Rome Urbi et Orbi. Despite the shadow of the temporal power of the Church of Rome and began a prodiguiso development scientifico.I scientific centers of Babylon, Alexandria, Athens are replaced by poles diToledo.Firenze, Wejl, Prague, Cambridge This dominated dalfondatore of celestial mechanics , the Newton (1643-1727)

ASTRONOMY OF tHE FOUNDING FATHERS

Grava the ominous shadow of the Holy Office. Although this is in particular the positional astronomy that soars into the sky. and is fatal clash with the spirituality of faith in dogmatic religion. The man wants to understand where he is in the immensity of creation, incredulous of the dogma that puts the Earth at the center of the Universe. Here is the man of destiny, in 19 centuries by **Hipparchus.**

One unknown. This N.Kopernig (1473-1543), born Thorm (Poland), studied in Krakow
And then at the University of Bologna (1496)
following the courses of theology and astronomy.
In 1501è nominaio canon Frauenburg
Si stops in Italy. In Padua, Ferrara, where he studied medicine and canon law.
From 1504 he settled permanently in Emerlamd (East Prussia) where, mostly, dealing with astronomy

In 1514 Laternense approached by the Council on the phases of the moon, not being sure of the position of the stars silent
Earth-Moon-Sun (STO2 Fig4) after numerous Earth-based observations with Mars notes that the Sun is stationary.
But for fear of Roma Caput
Urbi et Orbi not by the
announcement "SOL STAT"

Sol Stat è un uragano che demolisce il dogma della scolastica per cui il pianeta **Terra** , dimora dell'uomo immagine di Cristo, era il centro del cosmo e su tale linea si sono succedute opere di carattere speculativo e religioso carenti di valore A partire da **Ipparco**(pg-4) era stata individuata la distanza della Luna rispetto alla **Terra**, Fig5(a) Allora detta TA la distanza **Apogea** e TP la **Perigea**. Detto R il raggio del cerchio e detta c la distanza dal centro C **Ipparco** era riuscito a determinare la distanza della Luna dalla **Terra**. Gli astronomi dell'epoca con questo metodo sono riusciti a calcolare le posizioni di altri astri come la Luna e il Sole - **Mercurio-Venere-Marte -Saturno** Nel 2° secolo(d.C.)appare una ricerca con osservazioni planetarie, definta dagli astronomi << **il sistema matematico del mondo**>> Il testo " **Almagesto** dall'Arabo All Gabr) scritto dal greco **Claudio Tolomeo** : Secondo il quale , l'universo,Fig 5b,ha per centro la **Terra** immobile. Tutti gli altri astri in moto circolare uniforme (**significa che l'astro descrive archi circolari uguali in tempi uguali**) che si muovono fra il Perielio e l'Afelio . Il Sole sul cerchio deferente a ruotare intorno alla **Terra** La concezione tolemaica del creato è coerente con la intepretazione della filosofia scolastica , fatta propria dalla chiesa di Roma con conseguenze storico religiose- scientifiche secolari di grande importanza La visione astronomica dell'Universo ha inizio con il con il primo astrofisico della storia Al secolo il Greco **Ipparco di Nicea** immigrato alla corte dei Faraoni del tardo impero .Autore della famosa relazione spazio -tempo . Precisamente,Fig5a : $\frac{TP}{TA}=\frac{R-c}{R+c}=\frac{\beta}{\alpha}$ (1) ....

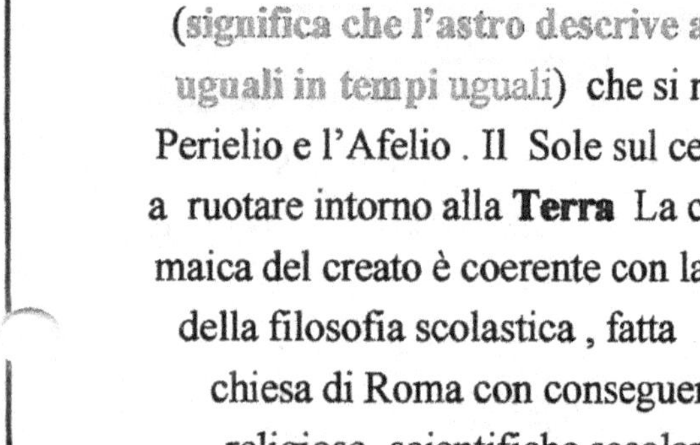

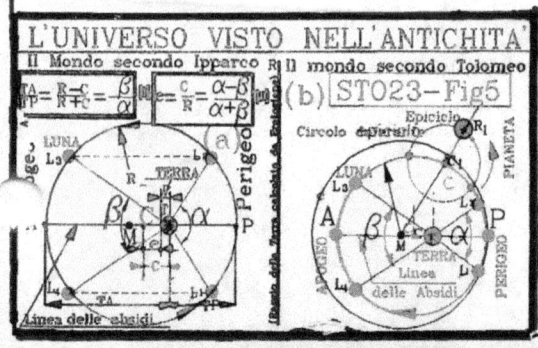

Sol Stat is a hurricane that demolishes the dogma of t school to which the planet Earth, home of man the im of Christ, was the center of the universe and on the li have followed the works of a speculative nature and lacking in religious value From **Hipparchus** (pg-4) identified an effective distance of Moon from the **Earth**, Fig5 (a)

Then the distance that **TA** and **TP** **Apogea** the Pe-rigea. Said R the radius of the circle and that c is t distance from the center C **Hipparchus** was able to determine the distance of the Moon from **Earth**. Th astronomidell'epoca with this method were able to calculate the posizionidi other celestial bodies like th Moon and the Sun - **Mercury-Venus-Mars-Saturn** In t 2nd century (d.C.) is a search with planetary observations, astronomers definta from the I << **mathematical system in the world** >> the text "**Almagest** from Arabic All Gabr) greek written by **Claudius Ptolemy**: According to which, the universe Fig 5b, has as its center the immobile **Earth** All the ot stars in uniform circular motion (**which means that t star describes circular arcs around the orbit**) that move between perihelion and aphelion with Soon in circle deferent to rotate around the **Earth** the Ptolema conception of creation is consistent with the interpretation of the scholastic philosophy, endorsed b the church of Rome with the historical consequences religious and scientific importance secolaridi the astronomical worldview of the Universe begins with t astrophysicist with the first century of history at the **Greek Hipparchus** of Nicea immigrant to the court c the Pharaohs of the late empire. Author of the famou relationship space - time. Precisely, Fig5a

$$\frac{TP}{TA}=\frac{R-c}{R+c}=\frac{\beta}{\alpha} \quad (1)$$

..: ....

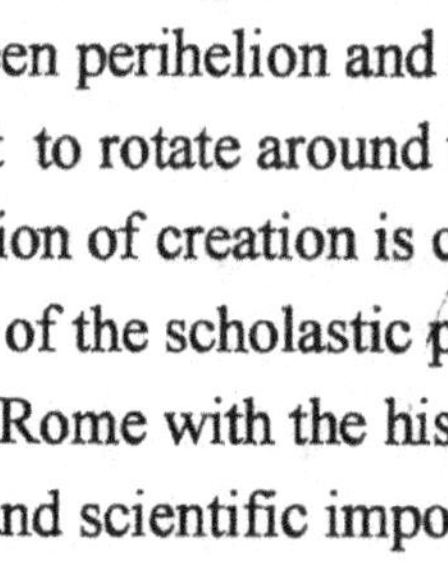
IPPARCO 190-120 a.c.

La relazione spazio -tempo per la misura della distanza **Terra-Luna** effettuata da **Ipparco** , Fig5a lega il rapporto degli angoli α e β : $\frac{TP}{TA} = \frac{R-c}{R+c} = \frac{\beta}{\alpha}$ [II] misurati dalla Terra al passagio della luna al perigeo ed all'apogeo . Il problema era allora di ardua soluzione dato che ,Fig3 per risolvere il triangolo **ABC** occore conoscere la parallasse (angolo dell'arco descritto dalla Luna ,fra **Perigeo** e Apogeo) .Rilevato da **Ipparco** del valore di 0° 54' (per l'epoca si disponeva di (sestanti primordilali) però con lo stesso principio di rifles-sione non di Fig27

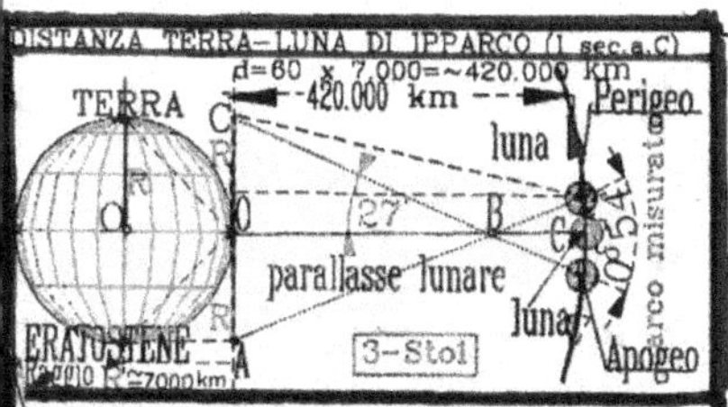

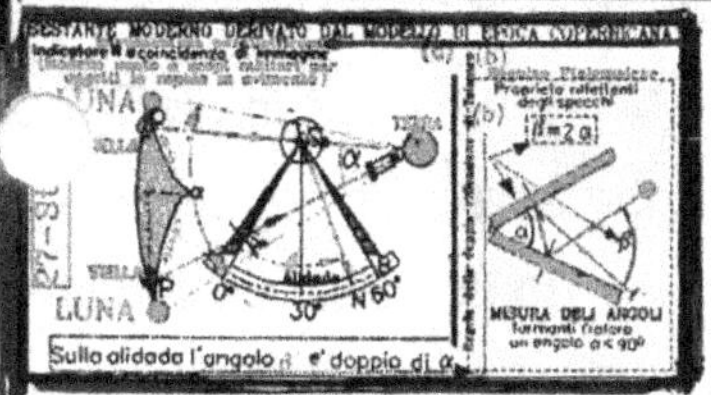

Ipparco conosceva il raggio R della Terra di **Eratostene** . Non si sa in che modo preciso anche perchè le funzioni trigonometriche erano ignote. Si sappeva che, per angoli piccoli , il seno è pari al suo arco misurato in radianti. La distanza TERRA -LUNAper l'epoca stata: $d_{TL} \cong 410.000$ km Oggi con il Radar è stimata km 379.768 km. Il valore ottenuto da **Ipparco** dedotto da una successione di osservazioni $\alpha_1 ,\alpha_2,...., \alpha_n$ di valori perigei ,Fig-27, ed apogei, $\beta_1,\beta_2,...,\beta_n$ Dai quali dedusse l'angolo di parallasse . Il valore medio 0° 54' ha reso possibile la determinazione d_{TL} Fig 3-Sto1 Ma per risolvere il Δ **OAB** **Ipparco** ha dovuto usare il lato AB = R(raggio dello sferoide terrestre) Si tenga presente che con i valori dei lati in gioco la **Terra** e la **Luna** vanno pensati puntiformi e quindi $OC \cong OB$ Ma è stato necessario comunque di usare il Raggio terrrestre R di **Eratostene** . Un altro Greco alla corte dei Faraoni Immigrato in Egitto divene bibliotecario della più famosa raccolta di manoscritti storici scientifico oggi dichiarata patrimonio mondiale dall'Unesco

The space-time relationship for measuring the **Earth-Moon** distance made by **Hipparchus**, Fig5a links the relationship of the angles α and β: == [II] measured from Earth to the passage of the moon at perigee and all 'heyday. The problem was difficult to solve since then, Fig3 to solve the triangle **ABC** is necessary to know the parallax (the angle of the arc described by the-**Moon,** between **perigee** and **apogee**).

Detected by **Hipparchus** of the value of 0° 54 '(for' age was available (sextants primordilali), but with the same principle of reflection not of Fig27

Hipparchus knew the radius R of the **Earth** Eratosthenes, it is not known how accurate partly becaus the trigonometric functions were unknown. sappeva is that, small angles, the sine is equal to its arc measured in radians

Earth-the distance **Moon** per the time was:

< 410,000 km= d_{TL} >

Today with the **Radar** is estimated at 379,768 km

The value obtained by **Hipparchus** deduced from a sequence of observations α1, α2, ..., αn perigees of values, Fig-27, and apogees, β1, β2, ..., βn

Dai deduced which the angle of parallax. the average value 0h 54 'has made possible the determination d_{TL}Fig 3- STO1

But to solve the Δ **OAB**

Hipparchus had to use the side AB = R (radius of the spheroid **Earth**)

Note that the values of the sides involved the **Earth** and the **Moon** to be thought Point and then **OB** ≅OC

But it was necessary, however, to use the radius R of terrrestre **Eratosthenes.** Another Greek at the court of th Pharaohs in Egypt Immigrant divene librarian of the mos famous collection of historical manuscripts science today declared a World Heritage mondialed at Unesco

I MALA TEMPORA- SULLE CENERI DEI MARTIRI CRISTIANI SORGE ROMA URBI ED ORBI (la fede nel dogma) pg-27

Riprendiamo quanto lasciato in sospeso a pg-23 Nel XV e XVI secolo d.C.Roma è la capitale della Cristianità.

Dopo un percorso virtuoso il papato assume contro il libero pensiero, irrispettoso della verità rivelata una forma sempre più dura e persecutoria. (Disgressione Attualmente il nuovo dogma è il profitto L'inquisizione è sostutita dalla dagli eccessi pubblicitari. L'organo inquirente ucante è il denaro il Sat'Ufficio lo strapotere bancario.

Possiamo sbagliare nel crudo giudizio. Ma sappiamo che l'uomo è quell'animale capace del bene in piccole dosi e del male in quantità industriale. Talvolta pericoloso quando crea un branco Nella pg-28 abbiamo riportato il quadro con effigiati i grandi pensatori che hanno dato all'umanità talvolta il pensiero scientifico e le leggi che lo governano. In primis troviamo **R. Bacone** fautore della sperimentazione scientifica ,come unica via per conoscere i disegni della creazione.Cozzando con ciò contro la filosogia scolastica della Chiesa di Roma . Condannato per vilipendio alla verità rivelata ,15 anni di duro carcere ! Il secondo preso di mira è **Galilei**. Incriminato per vehemente eresia(sosteneva la centralià del Sole di **Copernico**) Per la scolastica era il dogma e per **C. Tolomeo** la legge geocentrica con la Terra ferma e il resto degli astri rotanti su cerchi deferenti .

Galilei condannato al carcere con la intercessione dei Medici di Firenze passare il resto della sua via in esilio perpetuo! Lo stesso **Keplero** più tardi deve accorrere a difenedere dalla accusa di stregoneria la madre. Questa accusa al solito portava al rogo purificatore. Sfortunato **G. Bruno,** filososo del libero pensiero. Nessuno potè salvare lo sventurato filosofo da quell'orrendo fuoco del rogo "purificatore" , anno Domine 1600!

The MALA TEMPORA-ON ASHES OF CHRISTIA MARTYRS RISES AND ROME URBI ORBI (faith the dogma) pg-27

We take what was left suspended pg-23

In the fifteenth and sixteenth centuries-dCRoma is the capital of Christendom.

After a virtuous assumed the papacy as against freedom of thought, irrespective of revealed truth a form of persecution and getting tougher. (Digression Currently the new dogma is profit The Inquisition is sostutita from the excess publicity. Ucante The investigating body is th money the Sat'Ufficio the excessive power transfer. We might be wrong in the raw judgment. But we know that man is that ' animal capable of good and evil in small doses in quantities indus-trial. sometimes dangerous when it creates a herd in pg-28 we reported the effi-gees framework with the great thinkers who have given all'umani-ing sometimes scientific thought and the laws that govern it. Firstly we find **R. Bacon** proponent of scientific experimentation, as the only way to get to know the designs of creazione.Cozzando with what filosogia against the school of the Church of Rome. sentenced for contempt of revealed truth, 15 years of hard prison! The second pre-so targeted is **Galilei.** Indicted for vehement heresy (the Sun of Centralia supported the **Copernicus** For the school was the dogma and **C. Ptolemy** geocentri with the law firm and the rest of the Earth of the stars rotating on rims deferens. **Galilei** sentenced to prison with intercessone the Medici of Florence to spend the res of his way in perpetual exile! **Kepler** himself later must come in to difenedere from the mother accused of witchcraft. This accusation usual por- tava at the stake purifier. Unlucky **G. Bruno**, filososo of free thought. Nobody could save the hapless philosopher from that horrible fire of the burning "purifier", Domine year 1600

GALILEO GALILEI

1564 – 1642

KEPLERO

1571 – 1630

PERCHE' LA CHIESA RIFIUTA SOL STAT pg-28b

Il **Sol Stat Fig1-Sto1**,nicchia(b)è stato per la Chiesa di Roma un vero uragano per il coinvolgimento della fisolosia scolasticafatta propria. Cerchiamo di capire perchè. Per la scolastica l'uomo fatto ad immagine di Dio , è stato predestinato ad abitare, nel breve suo breve trasit, la Terra,nicchia (a). Quindi non poteva che essere il centro della creazione.In questo C.Tolomeo , Fig2, astronomo greco dell'epoca che sulla base delle sue osservazioni(allora si dispo-neva del solo sestante capace di misurare angoli di oggetti in mo-vimento) escogitò il modello di **Fig2-Sto2**, cosidetto del cerchio equante. Al centro dell'universo tolemaico la Terra con gli astri , come il **Sole**, in rivoluzione su cerchi concentrici(**equanti**). Per spie-gare l'alternarsi del giorno e della notte gli astri si muovevano sul ripettivo **epiciclo**. Questa immagine priva di valenza astrofisica(Basti sapere che la Terra ha ben quattorici movimenti 14 movimenti fra i quali quattro molto evidenti, a cominciare dalla rivoluzione attorno al **Sole**, di rotazione attorno al suo asse polare, di precessione degli equinozi,della nutazione (parallasse per effetto trottola). Che **Sol Stat** urtasse i nervi scoperti della immagine aristotelica di Roma Urbi et Orbi appare inevitabile. Che detronizzare la Terra fosse conside-rata dalla inquisizione vehemente eresia e che al Santo Ufficio incombesse l'obbligo di purificare con il rogo i dissacratori appare cosa del tutto in accordo con le conoscenze fisiche dell'epoca anche se la pratica purificazione ci sembra appartenere al fanatismo . Infatti relitti come Bacone (16 anni di carcere), Galilei (in esilio perpetuo), Copernico (fuggiasco) .

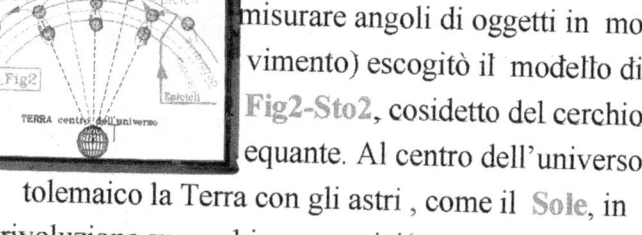

Ma alcuni scienziati hanno fatto di peggio con la costruzione della bomba A. Lanciata poi su Hiroshima e Nagasaki. In un istante 300.000 morti **fra cui bambini innocenti . VERGOGNA!**

WHY 'THE CHURCH REFUSES SOL STAT pg-28b

Sol Stat Fig1-STO1, niche (b) was to the Church of Rome a hurricane for the involvement of fisolosia scolasticafatta own. We try to understand why. For the school, the man made in the image of God, was destined to live in the short his brief trasit, Earth, niche (a). So that could not be the center of this creazione.In C.Tolomeo, Fig2, greek astronomer of the time that on the basis of his observations (then avail-neva the only sextant capable of measuring angles of objects in move-ment) devised the model of **Fig2-Sto2**, the so-called circle equant. At the center of the universe Ptolemaic Earth with the stars, like the **Sun**, in the revolution of concentric circles (**equants**). To explain the alternation of day and night the stars were moving on ripettivo **epicycle**. This free valence astrophysics (Suffice it to say that the Earth has very 14 movement Sun has fewer than 4 movements including four very obvious, beginning with the revolution around the Sun, the rotation around its polar axis, the precession of the equinoxes, the nutation (parallax effect for spinning).

Which **Sol Stat** ramming the nerves of the Aristotelian picture of Rome Urbi et Orbi seems inevitable. Which dethrone the Earth was regarded as heresy by the Inquisition and vehement that the Holy Office to lie with the obligation to purify with the burning of the desecration of the whole thing is in agreement with the physical knowledge of the time even if the purification practice there seems to belong to fanaticism. Indeed wrecks as Bacon(16 years in prison Galileo (in perpetual exile), Copernicus(fugitive)

But some scientists have done worse with the construction of the bomb A.

Then thrown on Hiroshima and Nagasaki.
In an instant 300,000 deaths

including innocent children SHAME!

PERCHE' LA CHIESA RIFIUTA SOL STAT pg-28b

Il Sol Stat Fig1-Sto1,nicchia(b)è stato per la Chiesa di Roma un vero uragano per il coinvolgimento della fisolosia scolasticafatta propria.

Cerchiamo di capire perchè. Per la scolastica l'uomo fatto ad immagine di Dio , è stato predestinato ad abitare, nel breve suo breve trasit, la Terra,nicchia (a). Quindi non poteva che essere il centro della creazione.In questo C.Tolomeo , Fig2, astronomo greco dell'epoca che sulla base delle sue osservazioni(allora si dispo-neva del solo sestante capace di misurare angoli di oggetti in mo-vimento) escogitò il modello di Fig2-Sto2, cosidetto del cerchio equante. Al centro dell'universo tolemaico la Terra con gli astri , come il Sole, in rivoluzione su cerchi concentrici(equanti). Per spie-gare l'alternarsi del giorno e della notte gli astri si muovevano sul ripettivo epiciclo. Questa immagine priva di valenza astrofisica(Basti sapere che la Terra ha ben quattorici movimenti 14 movimenti fra i quali quattro molto evidenti, a cominciare dalla rivoluzione attorno al Sole, di rotazione attorno al suo asse polare, di precessione degli equinozi,della nutazione (parallasse per effetto trottola). Che Sol Stat urtasse i nervi scoperti della immagine aristotelica di Roma Urbi et Orbi appare inevitabile. Che detronizzare la Terra fosse conside-rata dalla inquisizione vehemente eresia e che al Santo Ufficio incombesse l'obbligo di purificare con il rogo i dissacratori appare cosa del tutto in accordo con le conoscenze fisiche dell'epoca anche se la pratica purificazione ci sembra appartenere al fanatismo . Infatti relitti come Bacone (16 anni di carcere), Galilei (in esilio perpetuo), Copernico (fuggiasco) .

Ma alcuni scienziati hanno fatto di peggio con la costruzione della bomba A. Lanciata poi su Hiroshima e Nagasaki. In un istante 300.000 morti **fra cui bambini innocenti . VERGOGNA!**

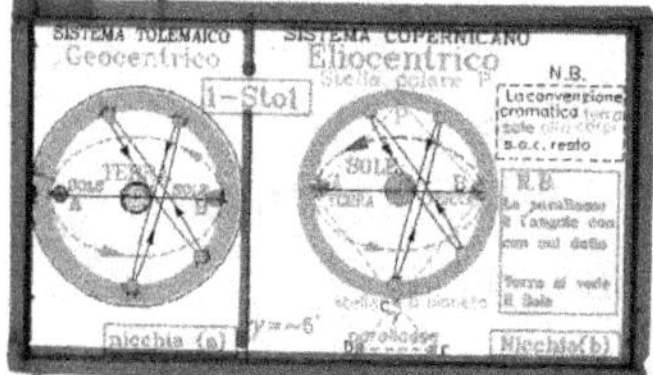

WHY 'THE CHURCH REFUSES SOL STAT pg-28

Sol Stat Fig1-STO1, niche (b) was to the Church of Rome a hurricane for the involvement of fisolosia scolasticafatta own. We try to understand why. For the school, the man made in the image of God, was destined to live in the short his brief trasit, Earth, niche (a). So that could not be the center of this creazione.In C.Tolomeo, Fig2, greek astronomer of the time that on the basis of his observations (then avail-neva the only sextant capable of measuring angles of objects in move-ment) devised the model of Fig2-Sto2, the so-called circle equant. At the center of the universe Ptolemaic Earth with the stars, like the Sun, in the revolution of concentric circles (equants). To explain the alternation of day and night the stars were moving on ripettivo epicycle. This free valence astrophysics (Suffice it to say that the Earth has very 14 movement Sun has fewer than 4 movements including four very obvious, beginning with the revolution around the Sun, the rotation around its polar axis, the precession of the equinoxes, the nutation (parallax effect for spinning).

Which Sol Stat ramming the nerves of the Aristotelian picture of Rome Urbi et Orbi seems inevitable. Which dethrone the Earth was regarded as heresy by the Inquisition and vehement that the Holy Office to lie with the obligation to purify with the burning of the desecration of the whole thing is in agreement with the physical knowledge of the time even if the purification practice there seems to belong to fanaticism. Indeed wrecks as Bacon(16 years in prison Galileo (in perpetual exile), Copernicus(fugitive)

But some scientists have done worse with the construction of the bomb A.

Then thrown on Hiroshima and Nagasaki.
In an instant 300,000 deaths
including innocent children SHAME!

STOROGRAFIA SCIENTIFICA # Sintesi riassuntiva delle precedenti pubblicazioi sotto indicate.Ordinate cronologicamente. Si ringrazia con l'occasione gli Atori dela presentazione

Anno 1960	: Satelliti artificiali ed orbite extraterrestri (Pres.ne :Prof.**Boaga** geodeta di Stato)
Anno 1975	: La fisica- matematica (Pres.ne : Ing **G.Montresor** (Pres.ne: Ord.Ingg di -VR)
Anno 1990	Centenario della relatività di Albert Einstein (Pres.ne :Ing Zenaglia -VI)
Anno 2000	: Dai Fasti tecnologici ai nefasti del potere per il potere(Pres,ne : Ing. **Poli**-VR)
Anno 2002	Le leggi del sistema solare di Keplero (Pre.ne Prof. E. **Burattini** Un. di VR)
Anno 2005	Astromia di posizione (Pre.ne Mons Piazzi PrefettoBiblioteca Capitolare-VR)
Anno 2007	Frammenti Scientifici del XXsecolo (Pres.ne dott Palazzo istituto aeronautico)
Anno 2010	Storiografia Scientifica (Vol. I) Patrocinata dal Comune di Verona)
Anno 2012	Storiografia Scientifica in corso d'opera Questo premesso.,allo scopo di dare

un senso della molteplicità degli argomenti che rientrano come radici nel presente Vol,VI Non si tratta dunque di un testo a livello narrativo.scientifico, nè di una raccolta stile enciclopedico (per questo bsta ed avanza Internet), Neppure di una pretesa teorematica di nuove leggi della fisica matematica,Ad es, quelle evocati nei frammenti scientific del XX secolo , da **Planck a Paul Dirac**, abbiamo associato come parte complementare ed integrante del formulario, dei grafici di supporto (disegmati con **Autocad LT97** e personalizzati)Ad es. nella Fig2-Stru1 , è stato indicato il cosidetto redshift $z=(\lambda_0-\lambda)/\lambda_0$ [I], con $\lambda=c/f$ [II] cioè :

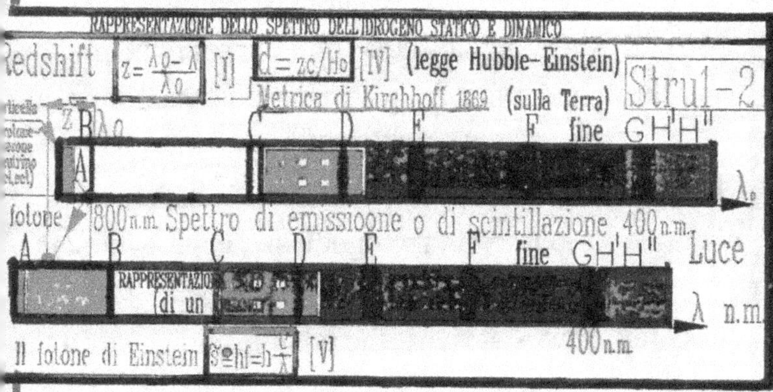

1-λ_0=lunghezza d'onda nel vuoto
2-λ = Idema nel mezzo materiale .
Consegue dalla [II] , se c=velocità della luce nel vuoto è una velocità limite costante che la $f= c/\lambda$ [III] cioè la frequenza f dell'onda diminuisce al al crescere di λ .Perfetto! Sorpresa finale. **Hubble** determinò un'altra costante Ho in suo onore. Ricordano la [I] potè costatare1che l'universo è in espansione determinando il Raddi pari a R=d=zc/Ho [IV] ,Meraviglioso ! ma una specie di masacro per le menningi La Fig 2 sintetizza per ilun l'ammasso stella quasar PKS agli estremi dell'universo il raggio R della universo in espanlse ento delle frequenze di una sorgente luminosa lontana dipende dalla sua velocità di allontanamento per il redshif[1] relazione **Hubble-Einstein** { d= zc/Ho) con Ho costante du **Hubble**. Per alleggerire il senso enigmatico della formiulazione ci è sembrto utile indicare laq frequenza di laboratori nello spettro λ_0 in metrica **Kirchhoff (as es dell' idrogeno isotopo trizio 1H1) ed unire con il redshift** in metrica Hubble Einstein di un astro o di un ammasso stellare. La freccia rivolta verso il basso rappresenta appunto la diminnuzione della frequenza dell'atomo appartenete al corpo celeste in fuga rispetto alla eclittica o se si vuole rrispetto alla Terra. Come mostra la rappresentazione grafica ,la z ha il vantaggio della visibilità della stessa I protagonisti sono i fotoni [V] di **Einstein**.Altro es. la parallasse associata al rivelatore della nicchia (a) con le modalità della osservazione integra quanto nella nicchia (d) di **Bessel** d= 1/p [II] delle distanze astronomi- che rilevate .Appare evidente il vantaggio della complementazione Per concludere Alea Jachta Est" l'intreccio fra storia e la scienza con deferenza dei testi sacri ." lo daremo a modo nostro Nella speranza di introdurre una disorganizzazione aperta alla fantasia . Audaces fortuna Juvat se .....Amen !

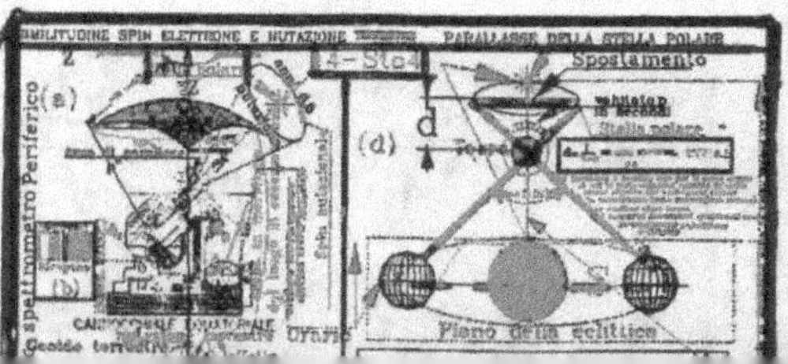

Friedrich Wilhelm Bessel in

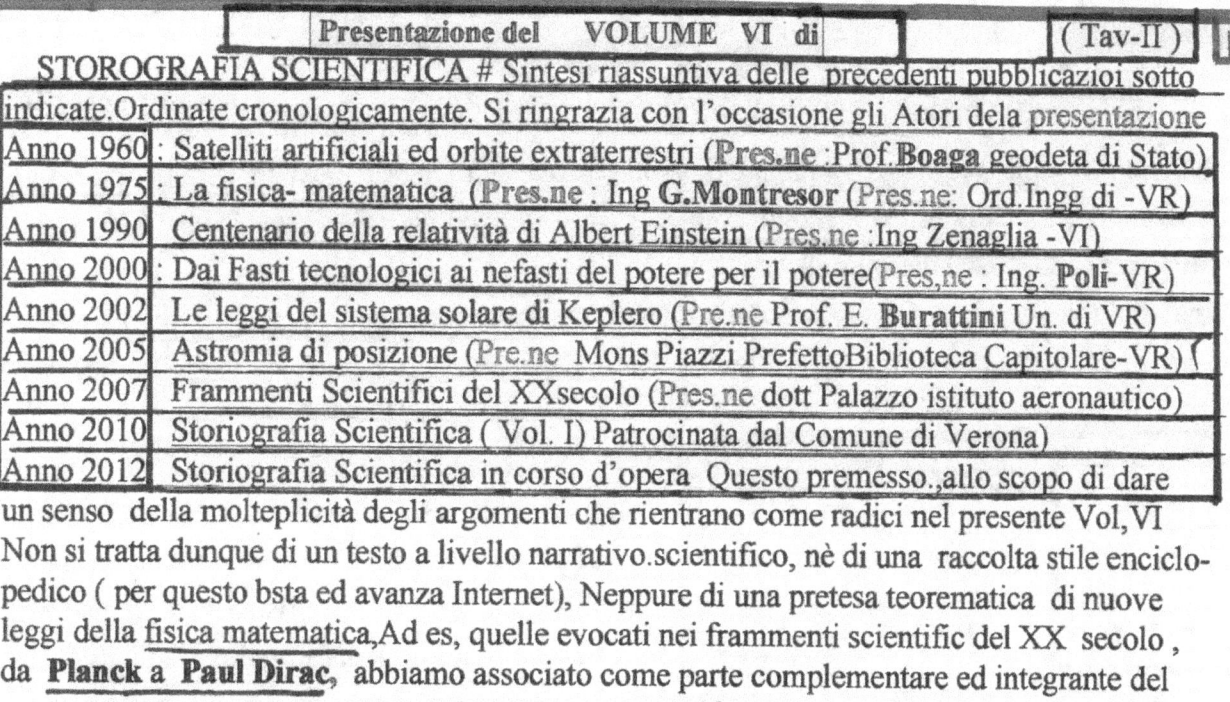

Ricordiamo brevemente le tre leggi dell'ottica geometrica riferita ai raggi luminosi in un mezzo materiale, riflettente e/o rifrangente, caratterizzata da alcune proprietà regolate dalla legge di **Villebrord Snell Van Royen**, latinizzato con **Snellius**(1580-1626) detto Snell.

Fisico, matematico, geodeta Olandese autore del teorema in concorso con Pothemot sulla

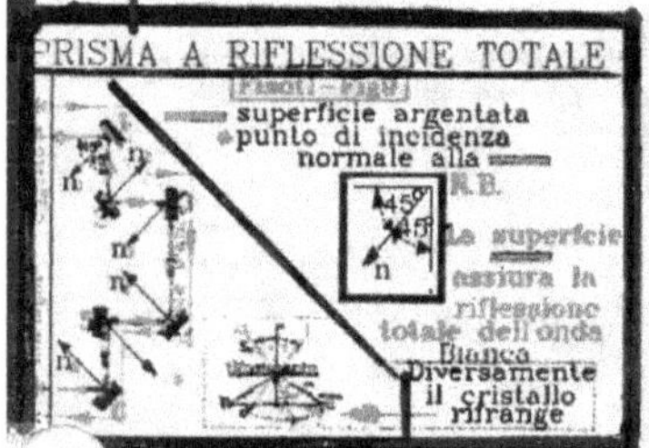

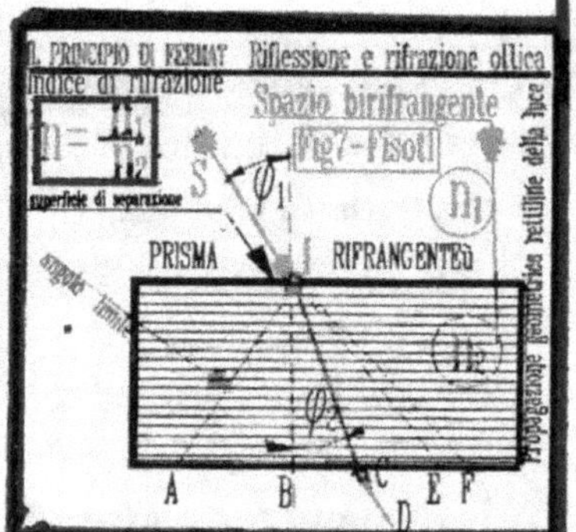

risoluzione di reti topografiche in applicazioni di trigonometria piana e sferica. In tali applicazioni si realizzano sistemi prismatici per il tracciamento di allineamenti rettilinei. Ciò dà la misura del principio di **Fermat** di **Fig-7** Un raggio di luce, che parte dalla sorgente **S** nel mezzo n_1 e incide sul prisma in **I**, per ci passa na normale con un angolo φ_1. Se la il prisma non è argentato e se $\varphi_1 > 0$ il raggio proseguirà il suo cammino nel mezzo n_2, avvicinandosi alla normale di un angolo $\varphi_2 < \varphi_1$

La legge di **Snell** afferma che il rapporto fra gli angoli dipende solo dagli indici di rifrazione dei mezzi attraversati. Postulata dalla equazione:

$$n_{12} = \frac{n_1}{n_2} = \frac{\sin\varphi_1}{\sin\varphi_2} \quad (1) \quad \text{con} \quad n_{12} = \text{Costante } (1_1).$$

Allora se si pone $\varphi_1 = \pi/2$ si trova la condizione :

$$\sin\varphi_2 = \frac{n_2}{n_1} = n_{21} \quad (2) \quad \text{con } n_{21} = \text{Costante } (2_1)$$

La(2) prova la invertibilità dei raggi di luce

Se la sorgente **S** è posta in n_2 la luce si riflette in **I** nello spazio fra i **raggi** per A e F come se la luce si riflette, **Fig3**, sulla superficie argentata

Let us briefly recall the three laws of geometrical optics referring to the rays of light in a material medium, reflective and / orifrangente, character-ized by some property under the laws of **Villebrord Snell Van Royen**, Latinized with **Snellius** (1580-1626) **Snell** said.

Physicist, mathematician, surveyor Dutch author of the theorem in competition with the Pothemot

to-polygraphic resolution of networks in applications of plane and spherical trigonometry. In such applications are realized prismatic systems for the tracking of alignments straights. This gives the measure of Fermat's principle in **Fig-7**

A ray of light, which starts from the source S in the middle n_1 and affect the prism **I**, passes na normal at angle φ_1.

If the prism is silver and if $\varphi_1 > 0$, the beam will continue its journey in the middle n2, approaching the normal angle $\varphi_2 < \varphi_1$

The **Snell's** law states that the relationship between the angles of refraction depends only dagliindici of the means through which . Postulated by the equation:

$$\mathbf{n}_{12} = \frac{n_1}{n_2} = \frac{\sin\varphi_1}{\sin\varphi_2} \quad (1) \text{ with } \mathbf{n}_{12} = \text{Constant } (1_1).$$

So if you put $\varphi_1 = \pi / 2$ is the condition:

$$\sin\varphi_2 = \frac{n_2}{n_1} = n_{21} \quad (2) \quad \text{con } n_{21} = \text{Costante } (2_1)$$

The (2) test the reversibility of light rays

If the source **S** is placed in the light n_2 is reflected in the space between the **rays** A and F as if the light is reflected, **Fig3**, on the silver surf

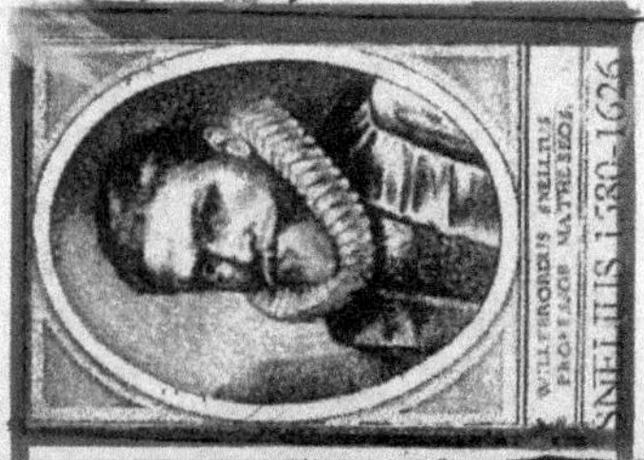

Premettiamo alcune note biografiche sul creatore del sistema solare formalizzato in tre leggi di cui la prima è supportata dalla Fig3-STO13

Johannes Kepler.

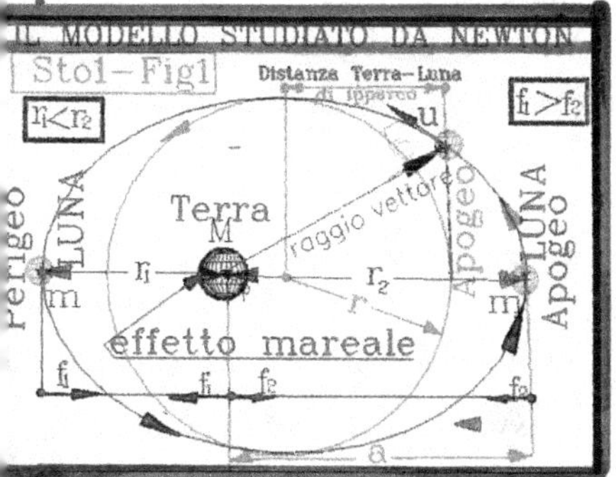

Nato a Weil (Wurtember) nel 1791.Dopo una dura infanzia causa salute e le modeste condizioni economiche della famiglia approda nel 1589 nel seminario di Tubinga dove incontra Mastli,fervido sostenitore di **Copernico** che lo inizia all'astronomia.Qui nel 1604 osservò una supernova . Divenne astronomo di Mattia II (Sigg. di Stiria e Carinzia) Infine successe a Tico Brahe alla direzione dell'Osservatori di Orjanebug. Predisse il transito di Mercurio la cui ombra si sarebbe proiettata l'anno come macchia solare. Per questo abbiamo ,nel caso della prima legge,parlato di magia del gigante Kepler. L'umanità gli deve in primis quello che è il primo sistema astronomico basato come amava precisare da leggi da cui dedurre la natura della creazione. Nel 1630 muore in Ratisbona .

Con l'uso del suo cannocchiale durante un decennio di ossservazioni, Fig3a , muta il sistema eliocen→trico(circolare) ,Fig 3b nel sistema in eccentrico (ellittico) della Terra e dei pianeti delle tre leggi.

Con queste Keplero ha offerto alla umani tà un punto di appoggio per esplorare l'universo .

Infatti la Fig 18 mostra il sistema planetario visto dalla **Terra** con 2 pianeti interni(Mercurio e Venere,e gli altri 6 .Punto di appoggio astrofisico

.Keplero doceet .E' il solo dei grandi sepolto in fossa comune .Sic transit ... gloria mudi!

Premise some biographical information about the creator of the solar system formalized in three laws of which the first is supported by Fig3-STO13

Johannes Kepler

was born in Weil (Wurtember) in 1791.Dopo a tough childhood due to health and the modest economic conditions of the family lands in 1589, the seminary of Tübingen where he meets Mastli, fervent supporter of

Copernicus who initiates him to astronomy. Observed here in 1604 a supernova. He became astronomer Matthias II (Mr of Styria and Carinthia) Finally succeeded Tico Brahe to the direction of dell'Osservatori Orjanebug. He predicted the transit of Mercury whose shadow would be projected year as a sunspot. This is why we, in the case of the first law, talked about the magic of the giant Kepler. Humanity owes him in the first place what is the first system based astronomical as he liked to be specified by law by which to deduce the nature of creation. In 1630 dies in Regensburg.

With the use of his telescope during a decade of Etna's, Fig3a, mute the system eliocen-trico (circular), Fig 3b in the system in eccentric (elliptical) of the Earth and planets of the three laws.

With these **Kepler** has offered humanity a base for exploring the universe.

In fact Fig 18 shows the planetary system seen from the

Earth with 2 inner planets (Mercury and Venerre) and the other 6.

Astrophysicist Point of support.

Kepler doceet. And 'the only one of the great buried in a common grave. Sic transit gloria mundi!

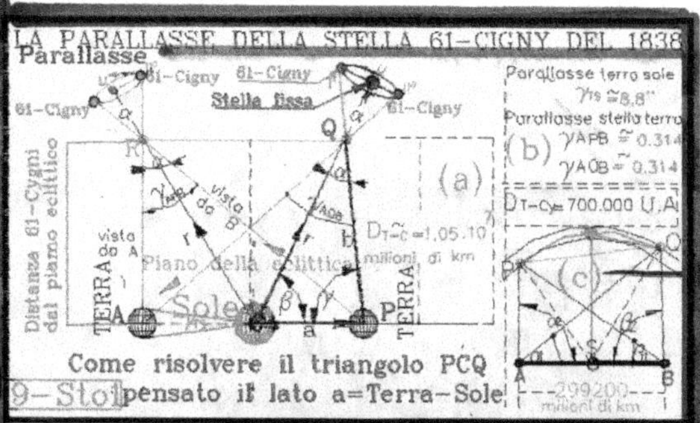

LA PRIMA LEGGE DI KEPLERO

Il pianeta scelto, Marte,è esterno alla Terra rispetto al Sole e siccome le orbite non sono complanari, ne circolari,ne omocentriche ,le equazioni legano, distanze e velocità angolari del raggio vettore SM sono precisi su scala siderale tale che i corpi in moto sianopuntiformi

P1- Definizione periodo di rivoluzione dei pianeti Se n è la velocità angolare della Terra e n' del pianeta esterno, si ha per l'intero periodo i rivoluzione attorno al Sole : $n-n'=\frac{360^n}{S}$ (1) . Relazione che vale pure per i due pianeti . Quindi si ha: $n-n'=\frac{360^n}{S}=\frac{360^n}{T}-\frac{360^n}{P}$ (2) . Quale trasformata si ha dividendo la(2) per 360°? <<Si ottiene la differenza della velocità angolare della Terra e Marte ad ogni ciclo di rivoluzione attor no al Sole>> $n-n'=\frac{1}{S}+\frac{1}{T}-\frac{1}{P}$ (3). Ora nel sistema eliocentrico il Sole è riferito al raggio vettore ST pianeta Terra e SP di altro pianeta. S essendo il polo del sistema eliocentrico se $n \neq n'$.Si hanno due casi. Se n>n' per la (3) $\frac{1}{S}+\frac{1}{T}=\frac{1}{P}$ (3')

Quindi : $\frac{1}{P}=\frac{1}{T}+\frac{1}{S}=\frac{S}{ST}+\frac{T}{ST}=\frac{T+S}{ST}$ (4)

Infine: $P=\frac{ST}{S-T}$ (5) del pianeta **Marte** scelto da **Keplero** per la eccentricià della sua orbita rispetto al Sole che ha una frequenza ciclica micore della Terra e questo valer gli altri 6 pianeti. Se poi n'>n allora : $P=\frac{ST}{S+T}$ (6) è il pianeta interno ad avere una frequenza ciclica maggiore della Terra e si tratta dei pianeti interni(Mercurio o Venere) Nonostante le ripetute osservazioni e complicatis simi calcoli giunse alla conclusione che i pianeti del sistema solare non collimavano col il sistema eliocentrioc di **Copernico**.Da qui anni di osser- vazione per tentativi fra i modelli orbitali allo scopo di avere la concordanza fra quelle e questi. Finalmente conclude annunciando di aver risolto il **Misterium Cosmograficum**

. <I pianeti non si spostano su orbite circolari ma ellittiche, di cui il Sole **occupa uno dei fuochi**>

N,B +sono per definizione frequenze cicliche

Fig17-STO13

2- Copernico

THE FIRST LAW OF KEPLER

The chosen planet, Mars, is external to the Earth than th Sun, and since the orbits are not coplanar, it circular, it Homocentric, the equations linking, distances and angula velocities of the radius vector of SM are accurate siderea scale such that the moving bodies are point-like.

P1-**Definition period of revolution of the planets**

If n is the angular velocity of the Earth en 'of the outer planet, one has for the entire period of revolution around the Sun: $n-n' = \frac{360^n}{S}$ (1). Report that also applies to the two planets. So we have: $n-n' = \frac{360^n}{S} = \frac{360^n}{T} - \frac{360^n}{P}$ (2). What has changed is the dividenc (2) to 360o? << You get the difference of the angular velocity of the Earth and Mars during each cycle of revolution around the Sun >> $n-n' = \frac{1}{S} + \frac{1}{T} - \frac{1}{P}$ (3). Now in heliocentric system the Sun is related to the radius vector of SP ST planet Earth and another planet . S being the hub of the heliocentric system if $n \neq n$ 'you can have two cases. If n> n 'for (3) $\frac{1}{S} + \frac{1}{T} = \frac{1}{P}$ (3')

So: $\frac{1}{P} = \frac{1}{T} + \frac{1}{S} = \frac{S}{ST} + \frac{T}{ST} = \frac{T+S}{ST}$ (4) So, finally: $P = \frac{ST}{S-T}$ (5) of the planet **Mars** chosen by **Kepler** for eccentric of its orbit relative to the Sun that has a frequency less than the cyclic earth and the other 6 planets this valer

If then n '> n then: $P = \frac{ST}{S+T}$ (6) is the inner planet to have a cyclic frequency greater than the earth and it is the inner planets (Mercury or Venus)

Despite repeated observations and complicatis - similar calculations came to the conclusion that the planets of the solar system did not fit with the system of Copernicus.Da here eliocentrioc years of observation attempts to between orbital patterns in order to have the correlation between those and these. Finally concludes by announcing that he had solved the Misterium Cosmograficum.

<The planets do not move on orbit circular but lliptical, with the Sun> Nb:

cyclic frequency T(Great),P(planetes) Fig17-STO13

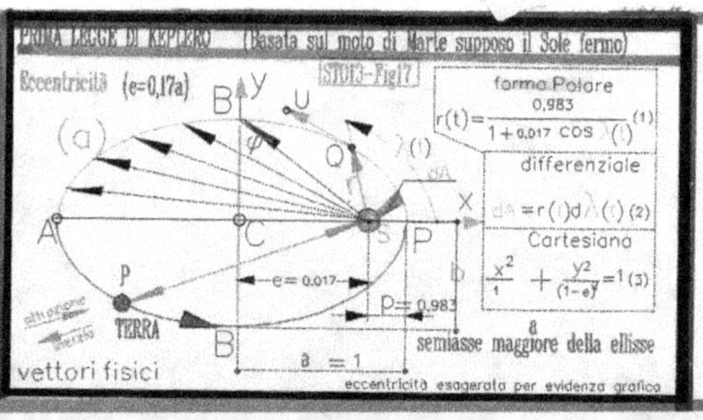

Conosce la teoria delle coniche sa che in generale ha la forma canonica algebrica di equazione: $a_{11}x_1^2+a_{12}x_2^2+a_{33}x_3^2=0$ (1) Riducibile con opportune trasformazioni alla forma, di **Fig3(b)** Infatti si ponga $x=\frac{x_1}{x_3}$ (1), $y=\frac{x_2}{x_3}$ (2) (c.c.o) e $a=\frac{a_{11}}{a_{13}}$, $b=\frac{a_{21}}{a_{13}}$ nella (1), con a>0 e b>0, Si ha:

$\frac{x^2}{a^2}+\frac{y^2}{a^2}=-1$ (3) Se si pone nella (1) espressa in c.c.omogenee: $a^2=-\frac{a_{23}}{a_{11}}$ (3), $b2=-\frac{a_{33}}{a_{12}}$ si ottiene la $\frac{x^2}{a^2}+\frac{y^2}{a^2}=+1$ (4) - Nella **Fig3** l'orbita dei pianeti è una ellisse con il Sole in uno dei fuochi (**la eccentricità è stata esagerata per ragioni grafiche**) e sovrapposto il cerchio copernicano. Si consideri ora il pianeta **P** in moto a velocità scalare **u**=Cost. con il **Sole** in **C**. Dalla fisica, nel moto circolare eliocentrico la energia del pianeta **P** $W_c=\frac{1}{2}mu^2=Cost.$ (5) (**conservativa**) Essa varia in regione inversa alla distanza r fra i baricentri. Si deve concludere che al variare del raggio vettore dipendente dall'angolo siderale ∂ varia il raggio vettore $\bar{r}(Q)$ Dal triangolo di lati a,b,c si ottengono le relazioni fra gli assi della ellisse. Quindi per le proprietà della coniche si trova la eccentricità cercata $e^2=a^2-b^2$ (6), $e=\frac{c}{a}=\frac{\sqrt{a^2-b^2}}{a}$ (7), $b=a-(\sqrt{(1-e^2})$ (8)

Keplero può dunque affermare di aver risolto il problema del **Misterium Cosmograficum**

I pianeti del sistema solare descrivono delle orbite ellittiche di semiassi ed eccentricità variabili

La condizione è cecessaria ma non è suffiente a caratterizzare le propriá cimematiche dei moti planetari Si noti che i pianeti del sistema copernicano si muovono tutti alla velocità scalare costante dipendente dal raggio r, come ?

knows the theory of conics knows that generation has th canonical form of the algebraic equatio: $:a_{11}x_1^2+a_{12}x_2^2+a_{33}x_3^2=0$ (1)

Shrink with appropriate changes to the shape of **Fig3 (b** In fact, it sets $x=\frac{x_1}{x_3}$ (1), $y=\frac{x_2}{x_3}$ (2) (cco) and

$a=\frac{a_{11}}{a_{13}}$, $b=\frac{a_{21}}{a_{13}}$ (1) with a> 0 and b> 0, one has:

$\frac{x^2}{a^2}+\frac{y^2}{a^2}=-1$ (3) If arises in (1) expressed in c.c.omogenee c.c.omogenee: $a2=-\frac{a_{23}}{a_{11}}$ (3), $b2-\frac{a_{33}}{a_{12}}$ you get the $\frac{x^2}{a^2}+\frac{y^2}{a^2}=+1$ (4) - In **Fig3** the orbit of the planets is an ellipse with the Sun at one focus (**the eccentricity has been exaggerated for graphic reasons**) and superimposed on th circle Copernican. Consider now the planet **P** moving with speed **u** = scalar Const. with the **Sun** in **C**. From physics, circular motion in the heliocentric energy of the planet **P**: $W_c=\frac{1}{2}mu^2=Cost.$ (5) (**conservative**) m It varies in inverse region at the distance r between the centers bari. It must be concluded that varying the radius vector dependent dall'agolo sidereal varies the radius vector $\bar{r}(Q)$ From the triangle of sides a, b, c we obtain the relations between the axes of the ellipse. So for the properties of the ellipse is the eccentricity sought $e^2=a^2-b^2$ (6), $e=\frac{c}{a}=\frac{\sqrt{a^2-b^2}}{a}$ (7), $b=a-(\sqrt{(1-e^2})$ (8)

Kepler can therefore claim to have solved the problem o Misterium Cosmograficum **The planets of the solar system you describe the orbital eccentricity and elliptical shafts variables** The condition necessary but non problem is solved is not suffiente to characterize it cimematiche of planetary motions

Note that the Copernican system of the planets all move at the speed Scaling constant dependent on the radius r, as?

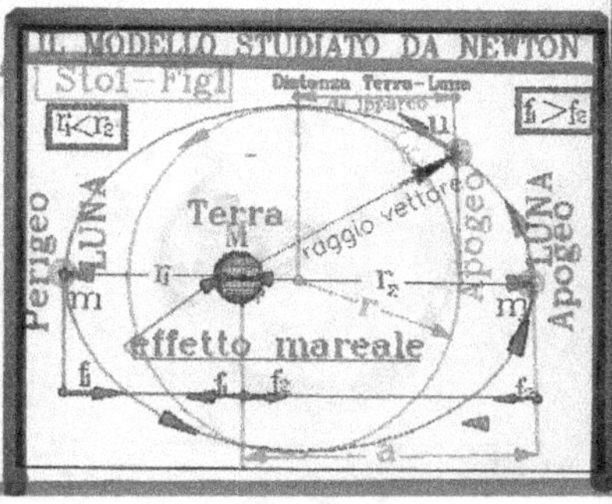

3- Keplero

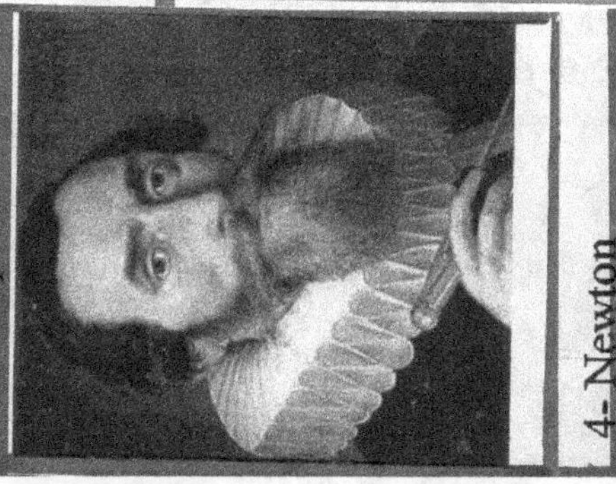

4- Newton

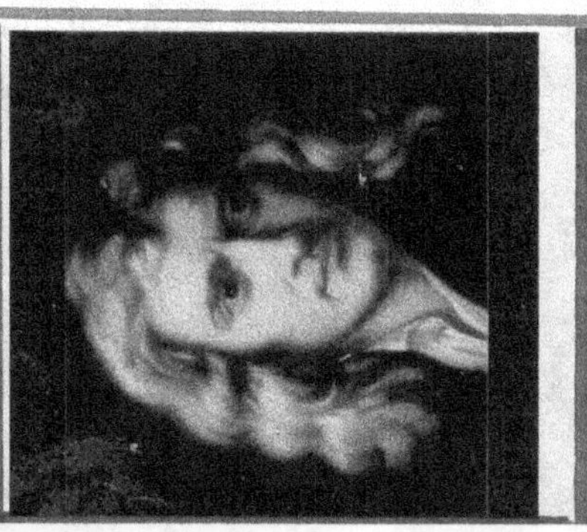

Se c'è un problema per risolvere il quale si deve far ricorso alla storia pregrersa e all'aiuto della tecnologia(allo scopo **Keplero** ha realizzato un cannocchiale telescopico di portata e canpo ni-gliore di quelli

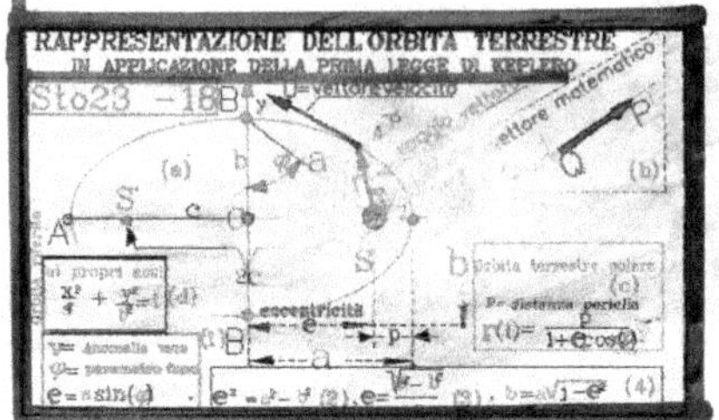

all'epoca in uso) .Che malgrado il contributo di tutti resta nondi meno incerto e ap-prossimato , queste sono appunto le tre leggi di **Keplero**.In altre parole si può dire che queste vanno intepretate con rigore entro i limiti assiomatici. Cioè vere fino a quando non si scopra che sono false.

La seconda legge di Keplero

A questo punto sappiamo per quantodetto in precedenza che i **pianeti P** su orbita circolare hanno velocità **u** scalare **costante** e tale è la energia . Quindi la energia cinetica di **P** è $\mathbf{Wc}=\frac{1}{2}m\,\mathbf{u}^2$ (1) , supposto che **P** orbiti sul cerchio di raggi r=1. Ricordiamo che r=1è(indipendente dalla unità di misura prescelta: .. cn,km,.) Se si suppone che **P** circoli su r≠1≠ 0 >0 la (1) sarà: $\mathbf{Wc}(r)=\frac{1}{2}mr\,\mathbf{u}^2$ (2) che, per u=Cost: aumenta e diminuisce con r . Questo era il mistero. Dalle osservazioni **Keplero** era giunto a conlusioni opposte ossia : $\mathbf{Wc}(r)=\frac{1}{r}\,m\,\mathbf{u}^2$ (3) , in base alla quale **l'energia cinetica diminuisce se r aumenta e viceversa**

Ipse dixit << Ciò che **Copernico** fece più per felice **intuizione** che per sicura **deduzione**>>

Keplero con la prima legge aveva stabilito che l'orbita descritta dai pianeti era ellittica con la seconda ed un complicato discorso analogo alla prima legge con considerazioni legate al sistema eliocentrico dimostra la 2^a legge che annunciamo

<< I **pianeti P** **non descrivono archi uguali in tempi uguali ma aree uguali in tempi uguali**>>

In sintesi: $\frac{dA}{dt}=\mathbf{Cost}$ (5) . Incompatibile con u=Cost, I nfatti da questa $\frac{du}{dt}=0$ Per la (5) si ha $\frac{du}{dt}=\mathbf{a}$ (6) chiameremo **a** il convitato ignoto

If there is a problem to solve which recourse must be ha to the history pregrersa and the help of technology (in order **Kepler** telescope has made a telescopic scope an canpo ni-gliore of those at the time use)

that despite the contribution of all remains none the les uncertain and ap-prossimato, these are precisely the thr laws of **Keplero**.In other words we can say that these a intepretate strictly within the limits axiomatic. That is tr until you discover that they are false.

The second law of Kepler

At this point we know for quantodetto earlier that the **planets** have **P** on a circular orbit velocity u and such i the constant scalar energy. Hence the kinetic energy of is $\mathbf{Wc}=\frac{1}{2}m\,\mathbf{u}^2$ (1), assuming that **P** on the circle o

orbiting radii r = 1. Recall that r = 1is (independent of th units chosen: .. m, km,.)

If we assume that P circles on 0 r1> 0 (1) will be: $\mathbf{Wc}(r)=\frac{1}{2}mr\,\mathbf{u}^2$ (2) that, for u = Cost: Increases and decreases with r. This was the mystery. Fromobserva-tions **Kepler** had come to some concluding remarks opposite ste ie: $\mathbf{Wc}(r)=\frac{1}{r}\,m\,\mathbf{u}^2$ (3) Toilet **whereby th kinetic energy decreases if r increases, and contrar**

Ipse dixit << What **Copernicus** was most happy to **intuition** that for sure **deduction**>> with **Kepler's** fir law had established that the orbit of the planets were elliptical with the second and a complicated discourse similar to the first law, considerations related to the heliocentric system shows that we state the 2nd law announcement

<< The **planetes P** **does not describe equal arcs in equal times but equal areas in equal times** >>

In summary: $\frac{dA}{dt}=\mathbf{Cost}$ (5) .

Unacceptatable for u = Cost,

I n fact this $\frac{du}{dt}=0$. The (5) for **Kepler is :**

$\frac{du}{dt}=\mathbf{a}$ (6) Call to **a** the convitato unknown

La seconda legge di Keplero

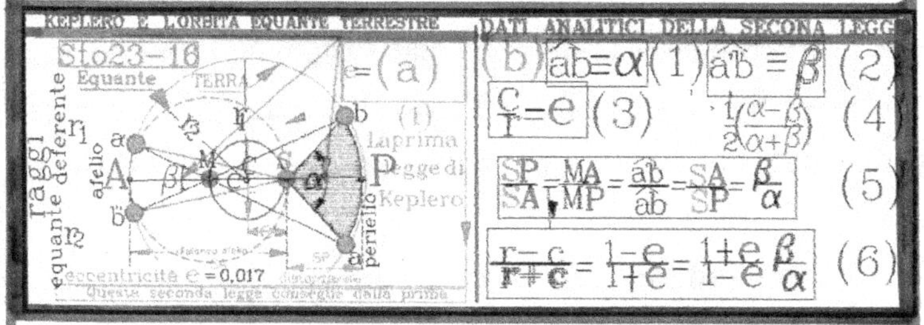

Keplero, introduce per l'orbita della **Terra** il cerchio **deferente** ,di raggi r_1 e centro C, **Fig 16(a)**, con il **Sole Ś** , del cerchio **equante** di raggio r_2, in accordo con la visione di **C.Tolomeo**(90-\160 d.C.) , autore dell'Almagesto, e di **Ipparco** (3° sec.a,c) già descritti.**Keplero** ,**Fig.16(b)** . Rifa il percoso storico nella formulistica **(b)** Evoluzione che ha richiesto secoli per le conclusioni. La legge delle aree è stata verificata da **Keplero** solo nei settori ellittici in corrispondenza al perigeo **P** ed apogeo **A** di eccentricità **e=0,017** per la Terra , oltre che per Mercurio e Venere. Per Marte ha trovato **e= 0,093** prescelto appunto per la grande eccentricità ,riferite in % al semiasse a Terra(**e**=0,017a),Marte(**e**=0,093),Plutone(**e**=0,248)

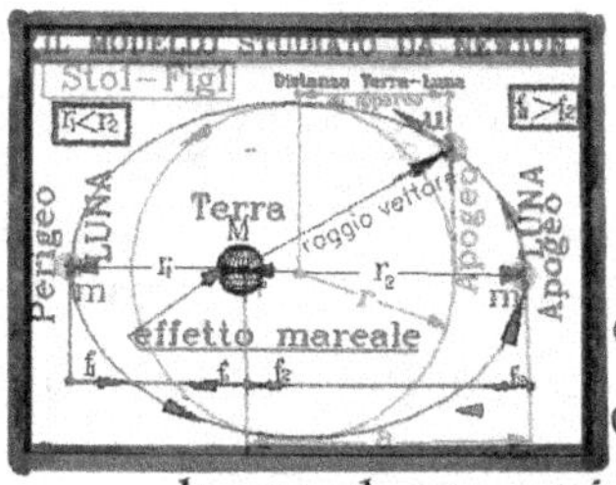

LA TERZA LEGGE DI KEPLERO

Le 2 leggi precedenti furono pubblicate nel 1609. In sintesi la prima afferma la divergenza delle orbite planetarie circolari e la seconda stabilisce che la velocità uniforme è sostuita dalla costanza della velocitè areale,al limite dA/dt= Cost. La terza legge si enuncia << **I quadrati dei tempi di rivoluzione stanno tra loro come i cubi dei semiassi delle rispettive orbite ellittiche**>>

Se a_1 e a_2 sono i semiassi di 2 pianeti dei 9 e indichiamo con P_1 e P_2 i periodi di rivoluzione : $a_1^3 : a_2^3 = P_1^2 : P_2^2$ (1) delle rispettive rivoluzione La **Fig27** con t≡ b lega i tempi di rivoluzione.....

The second law of **Kepler**, introduces for the **Earth's** orbit the rim **deferens**, rays r_1 and center C, **Fig 16 (a)**, with the **Sun S**, the circle **equant** of radius r_2, in accordance with the vision of C.**Tolomeo** (90 - \ 160 AD),the author of the Almagest,and H**ipparchus (third sec.a, c)** already descritti.**Keplero**, **Fig.16 (b).**

Rifa ritch town in formulisica **(b)**
Evolution be that required centuries to conclusions.

The law of areas has been verified by **Kepler** only in the areas elliptical corrispondens at perigee and apogee **P A** of eccentricity **e = 0.017** for the Earth, as well as Mercury and Venus.
For Mars has found **e = 0.093** chosen precisely because of the large eccentricity, in% related to the drives haft to Earth (**e** = 0.017 a), Mars (**e** = 0.093), Pluto (**e** = 0.248)

THE KEPLER'S THIRD LAW

The 2 previous laws were published in 1609.
In summary, the first states the divergence of the planetary orbits circular and the second states that the velocity is uniform sostuita by the constancy Speed Control range, the limit **dA / dt = Const**
The third law to enuncciate
<< **The squares of the times of revolution are among them as the cubes of the midaxis of their elliptical orbits** >>
If a^1 and a^2 are the semi-axes of 2 of the 9 planets and denote by P_1 and P_2 periods of revolution:

$$a_1^3 : a_2^3 = P_1^2 : P_2^2$$

of the respective revolution The **Fig27** tb alloy with the times of revolution ....

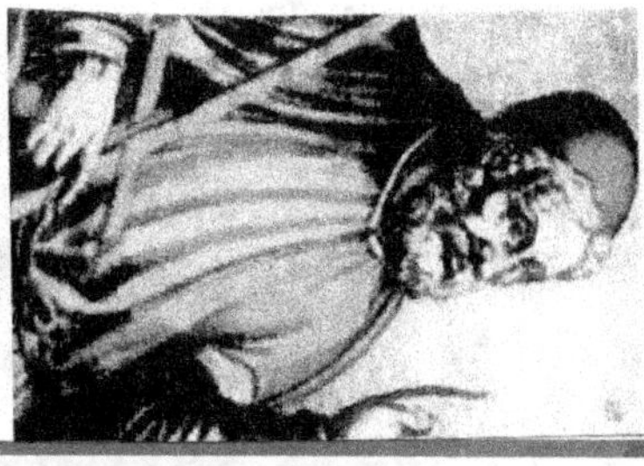

TOLOMEO

100 – 175 a.c.

Due giganti a confronto . Keplero parlando di Copernico e del sistema solare ortocentrico arriva a Sol Stat! più per felice intuizione che per sicura deduzione. E che le deduzioni sul sistema solare ecceentrico fossero esattamente dedotte da **Keplero** è un fatto certo , nei limiti e condizioni delle approssimazioni astronomiche

I.Newton (1643-1727) dirà perchè si muovono nelle loro predestinate orbite i pianeti del sistema Solare di **J.W.di Keplero** (1571-1630) che, con le 2 leggi planetarie ha predetto come si muovono **Newton**,nato quando **Keplero** era morto, conosce quanto all'epoca si sapeva. In particolare era allora noto il raggio R della terra di **Eratostene**(276--195 a.C.) ,valutato di ~7.000 km

Conosceva pure la distanza Terra Luna misurata da **Ipparco di Nicea** (3°secolo a.C) di ~400.000 km. Appare naturale che, coscendo la $\overline{g}$ (gravità terrestre di **Galilei**) se la causa delle orbite ellittiche dimostrato da **Keplero** non fosse del tipo $\overline{g}$

IL PROBLEMA DI NEWTON

E'possibile che il modello distanza Terra-Luna **di Ipparco** sia il punto di partenza, non solo storico della meccanica celeste di **Newton** .

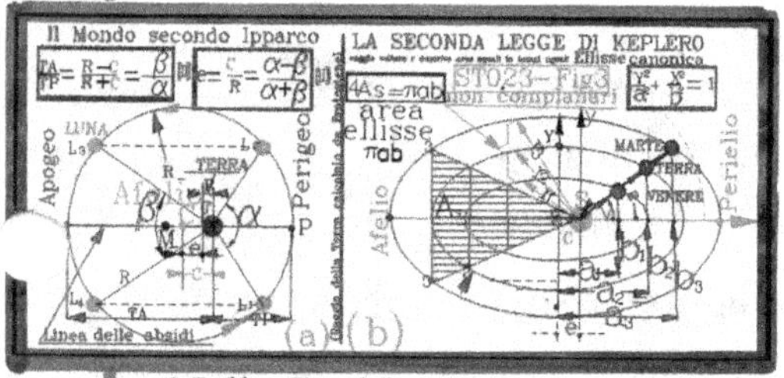

Nella Sto23- Fig3(a) è rappresentato il modello orbitale circolare usato da **Keplero** per il controllo delle osservazioni dirette in accordo con le fasi lunari di **Ipparco** che condusse al fallimento della ipotesi copernicana del modello ortocentrico del Sole Nella nicchia (b) il modello kepleriano della eccentricità del Sole rispetto ai pianeti in Orbite ellittiche in accordo con le osservazioni e la soluzione del **Misterium Cosnìmograficum**.

A questo punto è spontaaneo avere il sospetto che la causa sia riconducibile alla $\overline{g}$(Galilei).Come?

Two giants in comparison. **Copernicus** and **Kepler'** talking about the solar system ortocentric get to Sol Stat! more to intuition that for sure deduction. And th deductions about the solar system ecceentrico were exactly deduced from **Kepler** is a certain fact, within t limits and conditions of the approximations astronomic

I.Newton (1643-1727) tell you why they move into th predestined orbits of the planets of the Solar System **Kepler J.W** (1571-1630) who, with the 2 planetary lav predicted as they move **Newton, Kepler** was born whe he was dead, he knows what you knew at the time. In particular was the ra-known radius R of the earth by **Eratosthenes** (276-195 b.C.), estimated to ~ 7000 km knew well the Earth-Moon distance measured by **Hipparchus of Nicea** (3°Centiury b.C.) of ~ 400,000 k It seems natural, coscendo the (terrestrial gravity of **Galilei**) if the cause of the orbits elliptical policies show by **Kepler** was not the type $\overline{g}$

THE PROBLEM OF NEWTON

Spreader is that the model Earth-Moon distance of **Hipparchus** is the point of de-parture, not only histori of the celestial mechanics of **Newton**. In Sto23-Fig3 (a shows the circular orbital model used by **Kepler** to control of direct observations in accordance with the lunar phases of Hipparchus that led to the failure of the Copernican hypothesis of the model ortocentrico In the niche of the Sun (b) Keplerian model of the eccentricit of the Sun with respect to the planets in elliptical orbits i agreement with the observations and the solution of the Misterium Cosnìmograficum.

At this point is spontaaneo have the suspicion that

the cause is due to (Galilei). How?

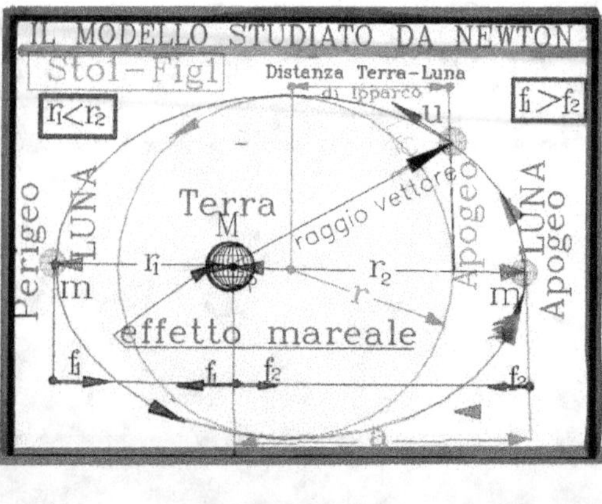

Chi fra i due giganti **Keplero.Newton** , dei due come aquila vola? La risposta la possono dare i poeti. Si possono solo confrontare le loro teorie e e rilevare la differenza fra le stesse e la inconfondibile diversità dei risultati. Le leggi di **Keplero** parlano di come si muovono gli astri la legge di Newton perchè si muovono . La loro vicenda umana è pure inconfondibile Alla sua morte a **Newton** sono concessi onori regali , e riservata sepoltura nella Abbazia di Westmister. Il Pope nella lapide scrive un meraviglioso epitaffio.

<< ISAACUS NEWTONUSQUAEM IMMORTALEM TESTANTUR TEMPUS, NATURA , COELUM : MORTALEM HOC MARMOR FATETUR. >> **<<NATURA AND NATURE'S LAW LAY HID IN NIGHT. GOOD SAID LET NEWTON BE! AND HALL WAS LIGHT >>**

<< Isaco Newton che il tempo,la natura e il cielo dichiarano immortale .Questo marmo lo dichiara mortale>> Newton meritevole e fortunato Newton ci ricorda con tristezza che l'uomo delle **tre leggi planetarie**, al secolo **Keplero** che, alla morte è stato sepolto in una fossa comune ed è forse l'unico grande dell'epoca del quale non ci sono i resti mortali . A pg-17 abbiamo parlato di **Newton** come uomo con luci ed ombre ,pur avendo poi addolcito il giudizio con molte più luci che ombre .Partendo da **Galilei ,** la cui legge di gravitazione terrestre per corpi pesanti in caduta libera sulla Terra, giunge con una felice idea ad interpretare l'effetto Terra come un caso particolare di materia operante per tutti i corpi di massa **m**$_1$ verso altri di massa **m**$_2$. Siano questi polveri o stelle. Come provarlo ? Questo è il problema di **Newton :** giustificare la legge :

$$F = K\left[\frac{Mm}{r^2}\right] \quad [1]$$

Una fatica, durata venti anni

Fallita al primo tentativo.

NEWTON EQUITE AURATO pg37

Who Keplero.Newton between the two giants, the two fly like an eagle? The answer can be given by the poets. It can only compare their theories ee detect the difference between same and the unmistakable diversity results. Kepler's laws speak of the stars as they move Newton's law because they move. Their human is also umistakable At his death in Newton royal honors are granted, and private burial in Westminster Abbey.

The Pope writes in a wonderful tombstone epitaph.

<< ISAACUS NEWTONUS QUAEM IMMORTALEM TESTANTUR TEM- PUS, NATURE, COELUM: MORTALEM HOC MARMOR FATETUR>>

<<. NATURE'S LAW AND NATURE HID LAY IN NIGHT. GOOD SAID LET NEWTON BE! HALL AND WAS LIGHT >>

<< Isaco Newton that time, nature and the heavens declare immortal.This marble to clear the deadly >>

Newton deserving and lucky reminds us, sadly, that the man of the **three planetary laws**, in the world **Kepler** that the death was buried in a mass grave and is perhaps the only great era in which there are the remains.

In the page-17 di Newton we spoke as a man with light and shadow, although then softened judgment with many more lu-us but shadows.

Starting from Galilei, whose law of gravitation terrestrial bodies for heavy drop-feed free on Earth, comes with a happy idea to interpret the effect of Earth as a particular case of matter acting on all bodies of mass **m**$_1$ to other mass m$_2$.

These are dust and stars. How to prove it?

This is the problem of **Newton** : justify the law:

$$F = K\left[\frac{Mm}{r^2}\right] \quad [1]$$

An effort, lasted twenty years.

Failed on the first attempt.

IL PROBLEMA DI NEWTON

La ossessione di **Newton** è quella di estendere la gravità terrestre di **Galilei** alle masse M pesanti dell'intero universo. Interpreta correttamente la interazione dell'effetto gravitazione sui corpi di massa m in moto libero rispetto alla **Terra** come caso particolare.Generalizza la proprietà alle masse M(dominanti supposte fisse) Formalmente la gravità galileiana è espressa dalla equazione vettoriale $\overline{F} = \overline{g}\frac{Mm}{r^2}$ (1) . La parte scalare $\frac{Mm}{r^2}$ e data dal prodotto delle masse supposte in uno spazio tridimensionale euclideo, definito dalla distanza **r** fra i baricentri di M e m. La r in c.c.o.vale $r = \pm\sqrt{x^2+y^2+z^2}$ .Consegue che la (1) è a simmetria sferica.Fisicamente significa che, supposta la **Terra M** sferica riducibile ad un punto g l'effetto attrattivo bivalente(M↔ m) nel punto g∉ M dipende dalla sola r

Il problema di Newton Se M ed m sono pianeti o stelle del firmamento la(1)diventa formalmente: $\overline{F} = \overline{k}\frac{Mm}{r^2}$ (2) con la $\overline{k}$ costante di attrazione universale. A prima vista sembra che **Newton** abbia semplicemente sotituito $\overline{g}$ con $\overline{k}$ Al contrario non sono gli esperimenti con il piano inclinato e le palline che scendono, **Fis1-Fig14**

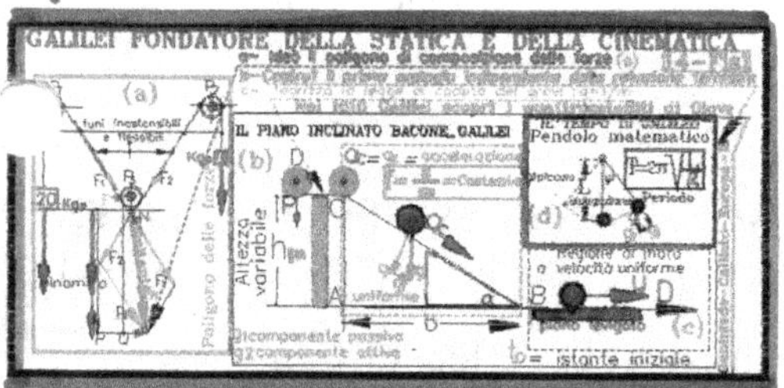

Galilei primo fra tutti ha definito , osservando le palline sul piano da B a **D** la **u** della pallina P è velocità uniforme misurando ,nicchia (d), il tempo t (**impiegato dalla pallina per pasare da B a D**)da cui la definizione<< **un corpo si muove di velocità scalare uniforme se percorre in tempi uguali distanze uguali**>>Va precisato che nel piano **O-B** inclinato agisce la (1) esclusa in **B-D**dalla reazione del piano levigato fisso alla terra.Ma la (2) postulata da **Newton** è diversa perchè mette in gioco tutte le masse cosmiche

THE PROBLEM NEWTON(Newtom problem

The obsession of **Newton** is to extend the Ehart gravi to **Galilei's** heavy M to . He Correctly interprets the gravitation effect on bodies with mass m in free moti reslated to the **Earth** as particular case.
He generalize the property to masse M (assumed as fixed dominantts)

Formally, the Galilean gravity is expressed by the vec equation $\overline{F} = \overline{g}\frac{Mm}{r^2}$ (1). The scalar part and given by t product of the masses, assumed in thre dimensional Euclidean space, de-fined by distance r between gravity of M and m. $r = \pm\sqrt{x^2+y^2+z^2}$ in c.c. Conseguently (1) is a spherical simmetry
In phisical terms it means that, assumed the Earth is reducible to apoint g, hte attractive phisics means that, suppository **Earth M** spherical reducible a point g the attractive effect bivalent (M↔ m) at poi M∉ g depends only r

The problem of Newton

If M and m are planets or stars in the sky the (1) formally becomes: $\overline{F} = \overline{k}\frac{Mm}{r^2}$ (2) where $\overline{k}$ the constant universal attraction.
At first glance it seems **Newton** simply replaced $\overline{g}$ w $\overline{k}$ is the gravitatio constant On the other hand they are experiments with inclined plane and coming dowm bal
Fis1-Fig14

Galilei first of all defined , observing balls on line fro **B** to **D** the **u** P-ball uniform speed by measuring niche (the time t(**taken by go from B of D**) hence the term << **body moves speed scalar uniform when equal distance in equal times along** >>
It should be noted that in plane OB acts (1) excluded B-D. from the to the ground
But.Ma (2) postulated by **Newton** is different because puts into paly all the cosmic masses.
This is the problem.

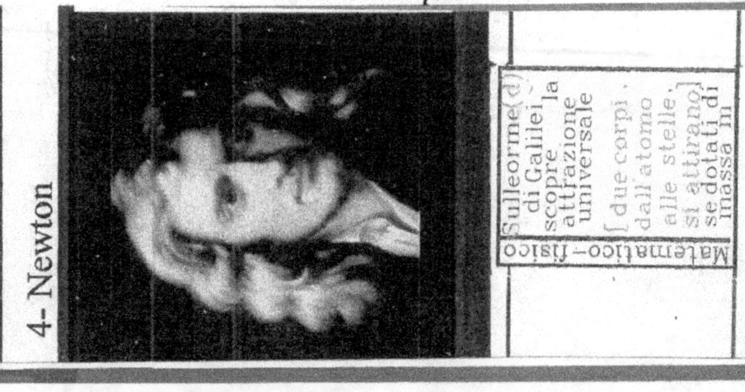

Newton con felice intuizione considera la prova sperimentale della forza $F=\bar{k}\frac{Mm}{r^2}$ (2) ,con la $\bar{k}$ costante di attrazione universale,riconducibile al modello dell'astrofisico **Ipparco** di Nicea(28O-A.c) .Questi è riuscito a misurare l'angolo di **parallasse** , **Fig2 Sto1**, arco α= 55' 15''.

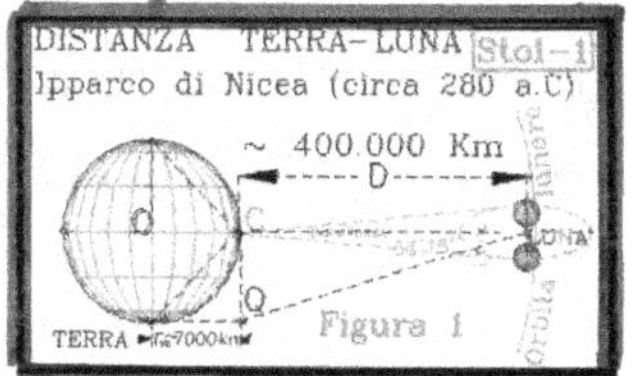

Usando il raggio terrestre **R** di **Eratostene** (**R**~ 7100 km) ha calcolato la distanza **Terra-Luna** .Da cui mediamente dedusse **d** $\cong$ 410.000 km . **Nel Vol."Dalla armonia della creazione..."** abbiamo riportato di **Newton** l'opera "**Philosophiae naturalis .Principiamathematica**" Scritta nel 1687 in cui appare la **equazione** $f_c=k^2Mm\,r^{-2}$ (1) sotto forma di forza applicata dalla Terra alla Luna ,equivalente alla (2). Forma più adatta ad esprimere le deduzioni a seguire . Allora se r_1 e la distanza **Terra-Luna** (al Perigeo), la (2) assume la forma $F_p=\bar{k}\frac{Mm}{r_1^2}$:(3) e ,all'Apogeo $F_a=\bar{k}\frac{Mm}{r_2^2}$ (4)

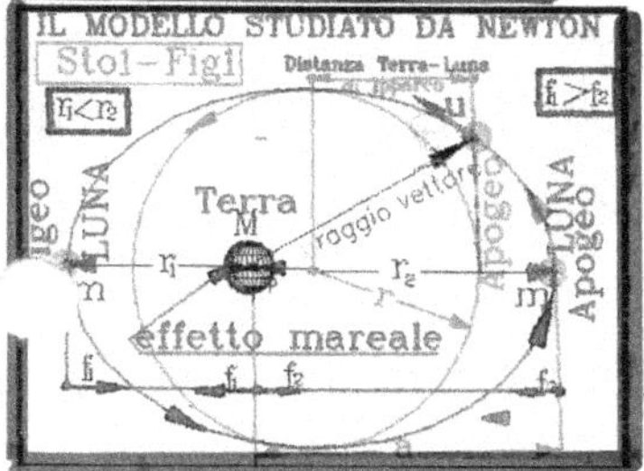

La (3) è la forza di attrazione Terra-Luna quando questa si prova al perigeo mentre la (4) quando la Luna si trova in apogeo La prova consiste nel fatto che se la Luna si trova al perigeo, sulla Terra **M** gli effetti si rivelano nell'alta marea.Viceversa per la(4)la bassa marea Eureka può aver pensato **Newton** . E' essenziale accordare l'effetto mareale con la invarianza di $\bar{k}$ ma questa, in predicato di diventare la costante di attrazione universale $\bar{k}$ variava con la posizione in orbita della Luna . Quindi la dimostrazione crolla e **Newton** si è procurato, forse l'esaurimento nervoso. Di certo una grande delusione poichè si tratta di un grande con una grande autostima. Addio giorni di gloria e dei castelli in aria.

Newton with intuition considers the experimental evidence of the force $\bar{k}\frac{Mm}{r^2}$ (2), with the constant of universal attraction, due to the model astrophysicist **Hipparchus** of Nicea (28O-Ac). Such was able to measure the parallax angle, **Fig2 STO1**, α=arc = 55' 15''.

Using the Earth's radius **R** of **Eratosthenes** (**R** ~ 7100 km) calculated the **Earth-Moon** distance. deduced from **d** $\cong$ 410,000 km. In Vol "**From the of creation** ..." we reported of **Newton**'s work " "**Philosophiae naturalis .. Principiamathematica**" [a] Written in 1687 in which the equation : $f_c = k^2Mm\,r^{-2}$ (1) in the form of the applied force from the Earth to the Moon, which is equivalent to (2).

Most suitable form to express deductions to falow Then if r_1 and the distance from the **Earth-Moon** system (at perigee), equation (2) takes the form :

$F_p=\bar{k}\frac{Mm}{r_1^2}$ (3) and Apogee : $F_a=\bar{k}\frac{Mm}{r_2^2}$ (4)

The (3) is the pull of Earth-moon at perigee when you try this while (4) when the Moon is at apogee the test consists in the fact that if the Moon is lo-cated at perigee, **M** the effects on **Earth** are revealed in high marea.Viceversa for (4) low tide Eureka may have thought **Newton.**

And 'essential to grant the tidal effect with the invariance $\bar{k}$ but this, in predicate become the constant of universal $\bar{k}$ attraction varied with the posi-tion in the orbit of the Moon.

So the proof collapses and **Newton** has procurat, perhaps a nervous breakdown.

Of course a big disappointment as it is a large with a great self-esteem.

He just had to do what I did. Place in the drawer with the elaborate goodbye to dreams of glory and castles in the air.

190 -120 a.c.

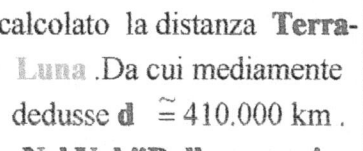

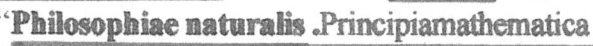

IPPARCO

Newton dopo aaver deposto nel cassetto la legge fallita : $\bar{f}=\bar{k}\frac{Mm}{r^2}$ (2) resta in attesa , segno che al sogno non aveva rinunciato. **Audaces fortuna juvat**. Infatti la fortu-na si presenta nel 1670 nella persona dell'Ing. geodeta **Jean Picard** (1620-1682) che misura un arco di geodetica ,da Sourdon a Malvoisine . La precedente geodetica era stata misurata da **Eratostene** circa 20 secoli prima.

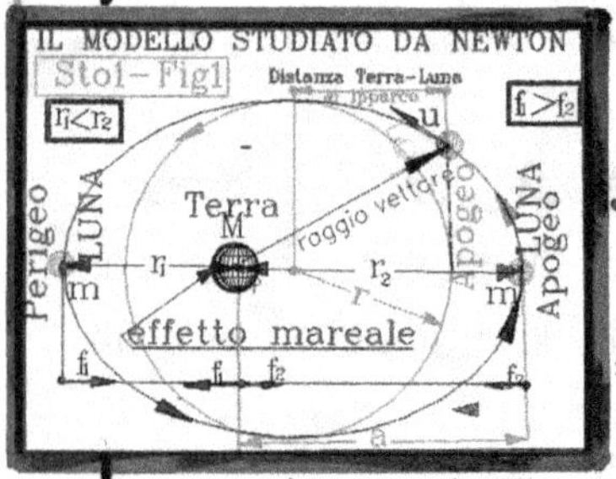

Infatti il raggio della terra, supposta sferica, è sta-to valutato di circa R= 7000 km.**Newton** : nota la r di Ipparco , Fig1 , assumendo la densità di massa media ρ della **Terra** (circa 1,1 volte H_2O) Valuta: $M=\rho V=\rho \frac{4}{3} p R^3$ (3) e la m con analogo procedimento(di larga approssimazione)

Avuta notizia della geodetica di **Picard** e intro-dotta nella (2) costata che l'effetto mareale LUNA-TERRA , sia al Perigeo che all'apogeo è in accordo di fase . Ecco finalmente! La(2) è la legge della **attrazione universale** .Valida sia per le poleveri che per i quasar (ammassi stellari) Il Il Pope nell'epitaffio sepolcrale di **IsaacNewton quirite aurato** scrive una meravigliosa sintesi

<< **Quem immortalem testantur tempus ,natura, Coelum .Mortalem hoc marmor Fathetur**>> ...

<<**Che il tempo la natura e il cielo attestano immortale.Questo marmo lo attesta mortale**>>

Se voi che leggete questa poesia non sentite alcu-na emozione, allora il nostro poeta vi dà un ultimo incoraggiamento << **Fatti non foste per viver come brutti ma per consegui virtute e conoscenza**!>> Diversamente<< **laciate ogni speranza voi che entrate** >> La natura ha creato l'animale uomo libero di scegliere il suo percorso e nessuno è immune da colpe , giudicato con il metro della morale. Ma c'è colpa e colpa ....

Newton placed in the drawer after aaver the law failed $\bar{f}=\bar{k}\frac{Mm}{r^2}$ (2) is waiting, a sign that the dream had no given up. **Audaces juvat luck**. In fact, the fortu-na occ in 1670 in the person of. surveyor **Jean Picard** (1620-1682) measuring an arc of geodesic, from Sourd to Malvoisine. The previous geodetic was measured b **Eratosthenes** about 20 centuries ago. In fact, the radi of the earth, supposed spherical, is being evaluated-t about R = 7000 km.

Newton: r known to **Hipparchus**, **Fig1**, taking the average mass density ρ of the **Earth** (approximately 1.1-f H2O).

Rate $M = V\rho = \rho\pi\frac{4}{3} R^3$ (3) and m by the same proced (large-approximation)

At the news of the geodesic **Picard** and intro-duced in finds that the effect of tidal MOON-EARTH, is at perigee tl apogee is agreement phase.

Here at last! Equation (2) is the law of **universal attraction**.

Valid for both poleveri that for quasars (star clusters) T burial of the Pope nell'epitaffio **Isaac Newton quirit aurato** wrote a wonderful summary

<< **Quem immortalem testantur tempus nature Coelum. Mortalem hoc marmor Fathetur** >> ...

<< **That time the nature and the sky attest immortale.Questo marble attests deadly >>**

If you who read this poem you hear AlCu-na emotion then our poet gives you one last encouragement

<< **You were not made to live like ugly but for the attainment of virtue and knowledge!** >>

<< **Otherwise just leave all hope ye who enter here**>

the nature has created the animal man free to choose his path and no one is immune from blame, judged by the yardstick of morality.

But there is blame and guilt ....

La storia di Roma è fin troppo nota per soffermarci in quello che è stato un impero esteso su tre continenti in termini temporali che con un certo arbitrio abbiamo fatto coincidere con la morte di Archimede(212 a.C.) e che la potenza e la ricchezza ha condotto allo sfacelo per la corruzione compagna inseparabile del potere.

Mentre i Greci elevarono la speculazione scientifica ai più alti livelli del sapere per il sapere il <Senatus Populosque Romanorum S.P.Q.R> Le colonie Greche della magna Grecia(Sicilia) e persino Atene divennero preda delle aquile latine si distinsero per rapacità e dispregio della filosofia del pensiero dando all'azione il dominio incontrastato in sintesi riassunto nel massimo dispregio per i valori spirituali che sono in sintesi il pollice verso contro il gladiatore sconfitto . Questa sintesi è riassunta con grande forza da **Cicerone** < In summo honore apud Graecos geomtria fuit,itaque nihil matematicis illustrius; at nos ratiocinandi metiendiqu utilitate hujiu artis terminavimus modum >Infatti Roma si ferma al conteggio con le dita delle mani :uns,duo,tres,...assolutamente impotente a costruire un contesto,anche se semplice strumento di una aritmetica primordiale.

La loro fede è l'utilità< del pensiero che guida l'azione volta a conseguire vantaggi a qualsiasi costo>. In certo qual modo la loro eredità è oggi palese nelle parole del nostro inno nazionale : La vittoria che schiava di Roma Iddio la creò!.

Ma torniamo a **Newton** . Il pensiero dominante era quello di scoprire le cause del fenomeno luminoso pensando alle stelle. Perciò infierendo la peste Cambridge ove insegnava. Attuò un eperirimento aprendo un forellino sulla imposta della casa paterna. Interponendo un prisma di Spato fluore.Ma la luce bianca attraversando il prisma dispiegava i colori Fig63 che chiamò **Spectrum**

The history of Rome is all too familiar to Soffer - rotten in what was an empire stretching across three continents in terms of time and with a certain arbitrariness we have made to coincide with the death of Archimedes (212 BC) and that the power and wealth has led to the debacle for corruption inseparable companion of power.

While the Greeks raised speculation scien-tific at the highest levels of knowledge to know the <Senatus Populosque Romanorum SPQR> The Greek colonies of Magna Grecia (Sicily) and even Athens became the prey of eagles Latin distinguished for rapacity and contempt of philosophy thought of giving to the dominion at a glance summary of the utmost contempt for the spiritual values which are in short thumbs down against the defeated gladiator . This synthesis is summarized with great force by **Cicero** < summo In honore fuit apud Graecos geomtria , itaque nihil matematicis illustrius , at nos ratiocinandi metiendiqu utilitate hujiu artis terminavimus modum > In fact, Rome stops to count with the fingers : uns , duo , tres , ... absolutely powerless to build a context , even if simple arithmetic means of a primordial . Their faith is the utility < thinking that guides action to achieve benefits at any cost > . In a way their legacy is evident today in the words of our national anthem : The victory that God has made a slave of Rome !

But back to **Newton**. The thinking was to find out the causes of the phenomenon of thinking about the bright stars .

So the plague raging Cambridge where he taught. Implemented a experiment by opening a small hole on the sets of his home .

Interposing a prism of Spat fluorescence , white light across the prism unfolded But the white light through he prism unfolded colors Fig63 he called **Spectrum**

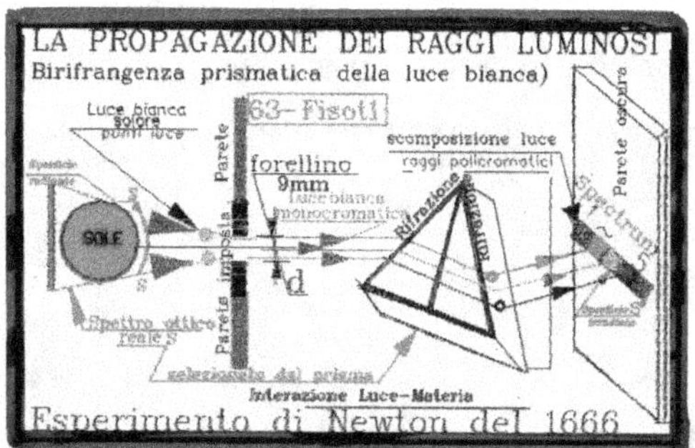

DA ROMA CAPUT MUNDI A ROMA URBI ET ORBI (Un nesso storico) pg-42

Nella pg-27 abbiamo parlato dei nefasti , secoli XV e XVI, in particolare noti come i mala tempora. Paradossalmente è il periodo d'oro del risveglio scientifico dell'Europa occidentale e che chiude l'epoca, a partire. secondo la nostra visio⟶ne con la caduta di Siracura ad opera del procon⟶sole romano **Marcello** (212 a.C.) . La storia ci racconta che un soldato romano, non riconoscendo ⟶ **Archimede** lo uccise .Poco credibile.Allora era già un famoso matematico . Autore della misura della semilunghezza della circonferenza di raggio unitario (π=3,141592654) come limite di due poligoni,**Fig6-Am1**, (r=1 poligoni di n_1 lati inscritto e n_2 lati circoscrito) quando si fa crescere $n_1$ed n_2.

Inventore del primo cannone idraulico detto Catapulta(datemi una leva e vi solleverò il mondo) Si è limitato invece a lanciare sassi misti ad olio bollente contro gli assalitori romani. Si può dedurre che **Marcello** ,espugnata Siracusa, abbia ordinato la sua esecuzione per le ustioni procurate ai suoi soldati. Nella storia scientifica la morte di **Archimede** segna lo spegnersi della luce diffusa dalla cultura speculativa greca.

L'ascesa di Roma si realizza in tutto il mondo estendendosi su tre continenti,Europa,Medio Oriente,Africa. I romani maestri nel dominare il mondo mediterraneo,sono degli eccellenti architetti.Collegano i continenti con le strade consolari e ponti ad arco statico ad arco reale. e sopra inseriscono aquedotti e viadotti dei quali sono presenti le vestigia disseminate ovunque

Hanno in discredito la scienza astratta, non posseggono un sistema di numerazione algebrico,i loro numeri aulici sono impotenti ad operare come operatori numerici La loro fede è

FROM ROME TO ROME CAPUT MUNDI URBI ET ORBI (A nesso storico) pg-42

In pg-27 we talked about the nefarious in the fifteenth and sixteenth centuries in particular mala tempora In corres pondence it is the golden period of the awakenin of Western scientific which closes the period, starting. according to our vision it with the fall of Siracura by the proconsum Roman **Marcello** (212 B.C). History tells us that a Roman soldier, not recognisind **Archimedes** killed him. Shortly credibile.Allora was already famous mathematician. Author of measuring length of half the circumference of unit radius (π = 3.141592654) as limit of two polygons, **Fig1-Am1**, (**r** = 1 sides inscribed polygons n_1 and n_2 sides circoscrito) when you grow $n_1$ed n_2.

Inventor the first cannon hydraulic said Catapult (Give me a lever and I will move the world) It is limited instead to throw stones mixed with boiling oil against the assailants Romans. It can be inferred that **Marcello**, to conquer Syracuse, has ordered his execution for burns inflicted to his soldiers.

In the history of science in the death of **Archimedes** marks the dimming of the light scattered by the speculative culture of ancient Greece.

The rise of Rome is realized in the world extending ove three continents, Europe, Middle East, Africa.

The Romans masters dominate the Mediterranean world, are excellent architetti

To conect continents with consular roads and bridges arched static arc real. and above to insert aqueducts anc viaducts which are the vestiges scattered everywhere

They discredited the abstract science, do not possess a numbering system algebrico , their numbers are powerless to operate as a courtly numeric operators thei faith is bread and circuses. Only gradiatore victorious deserves to vivere.Il won deserves death, decreestion with the thmb verso!

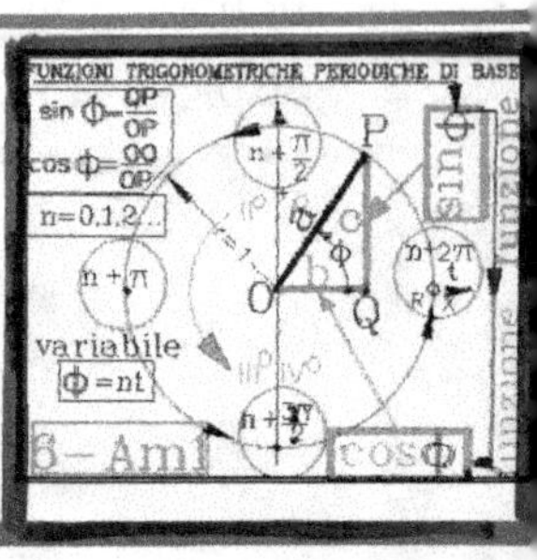

Chi è quel personaggio che va predicando la eguaglianza fra gli uomini ? Questa domanda è posta al S.P.Q.R.della Roma Caput mundi. La filosofia di **Cristo** è un pericolo per Roma e ordina alle sue falangi dislocate in Terra di Israel di imporre al Sinedrio di crocifiggere **Cristo.** Il Sinedrio è un potere disarmato ed è costretto ad ordinare la crocefissione di **Cristo** fra i ladroni perchè serva di monito.La storia di Cristo uomo nella fede ha lasciato una impronta più profonda e duratura di ogni altro nei popoli cristiani . Ai tempi di **Roma Caput mundi** era pericoloso manifestare sentimenti cristiani ma ciò servi a ingrossare le file degli apostoli e dei martiri. Lo sfacelo di Roma avviene con il concorso della potenza posseduta (una belva che divora se stessa creando ricchezza e corruzione. corruzione) con il concorso della fede che ha finito per dare nuova forma alla romanità cristiana Roma in breve diventa il centro urbi ed orbi. Ma in quanto umana è soggetta ,dopo la nascita e la crescita della potenza alla decadenza della fede. Qui prendiamo l'avvio,collegandoci ai personaggi del quadro iconografico di pg-28 (sintesi iconografica storiografica) Notiamo che il Solo **Newton** non ha subito violenza dalla inquisizione. Diversamente **R.Bacone e Galilei** sono stati incarcerati. Si deduce che il potere spirituale può mutare la fede in fanatismo. La prova è codificata nella sentenza del Santo Ufficio che manda sul rogo per la purificazione dei peccati il fisolofo **G.Bruno** da Nola.Arrestato dal Mocenigo a Venezia per il sospetto di stregoneria.Consegnato a **Roma Urbi ed Orbi**. Concluse la sua vicenda terrena affrontando la morte con dignità . Per **G.Brunoo le religioni sono forme di potere spirituale volte ad oscurare la realtà di Dio , che si identifica con la natura!Era una notte dell'anno Domine 1600**

Di quella Pira l'orrendo fuoco!

Who is that person who has been preaching the equality of men? This question is posed to S.P.Q.R.of the Roma Caput Mundi. The philosophy of **Christ** is a danger to Rome and ordered his phalanges located in the Land of Israel to impose the Sanhedrin to crucify **Christ**. The Sanhedrin is a power unarmed and was forced to order the crucifixion of **Christ** to the thief-ing it out as monito.La history of man in the faith of **Christ** has left a deep and lasting impression more than any other in the Christian peoples. At the time of **Rome Caput Mundi** was dangerous to express Christian sentiments but what servants to swell the ranks of the apostles and martyrs. The collapse of Rome is done with the course-with the power held (a beast that devours itself by creating wealth and corruption. Corruption) with the help of faith that has fi-nite to give new shape to the Roman Christian Rome soon became the center urbi and orbi. But as a human subject, after the birth and growth of power to the decline of faith. Here we take the start, connecting us to charac-gi's iconographic picture of pg-28 (iconographic synthesis historiographical) We note that the only **Newton** has not been subjected to violence by inquisition. Unlike **R.Bacone and Galileo** were imprisoned. We deduce that the spiritual power can change the faith in fanaticism. The proof is encoded in the judgment of the Holy Office, which sends the stake for the purification of sins by the fisolofo G.Bruno Nola.Arrest from Mocenigo in Venice on suspicion of stregoneria.Delivered **Urbi and Orbi** in Rome. He ended his earthly facing death with dignity.

For **G.Brunoo religions are forms of spiritual power-rare times to obscure the reality of God, which is identified with nature was a night of the year 1600 Domine**

Of that Pira the horrible fire!

GIORDANO BRUNO

1548-1600

Galilei nel dialogo sui massimi sistemi(1616), cui parteciano tre personaggi d'autore. Sagredo(lo stesso Galilei),Salviati(un architetto suo amico),Simplicius(Papa Urbano VIII) scrive :

<Simplicius> Io vi confesso che tutta questa notte sono andato ruminando le cose di ieri e veramente trovo di molto belle nuove, e gagliarde considerazioni ; con tutto ciò mi sento più stringer dall'oscurità di tanti grandi scrittori ed in particolare.....Voi scotete Sig.Sagredo....e sogghignate,.... come se io dicessi qualche esorbitanza.

<Sagredo> Io sogghigno solamente, ma crediatemi che io scoppio nel voler far forza di ritener le risa maggiori per un fatto che mi è sovvenuto.

<Salviati> Sarà bene che ci raccontiate il fatto a ciò forse il Sig.Simplicio non continuasse a credere di avervi esso mosso dalle risa>

<Sagredo> Non sarà mai al sicuro,come abbiano simili contradditori ma questo Sig.Simplicius non diminuisce la stravaganza del peripatetico(Aristotele).Il quale contro a si sensata esperienza(allude al Sol Stat copernicano) non produsse,allora, non produsse altra esperienza o ragioni, ma la sola autorità del puro : Ipse dixit!

<Simplicius> Aristotele non si è acquistato si grande autorità se non per la forza delle sue dimostrazioni e per la profondità dei suoi discorsi....

Da questo monologo sui massimi sistemi sembra poter concludere che Urbano VIII avesse molta considerazioni per il suo interlocutore Galilei pur non accettando il Sol Stat dallo stesso propugnato. Sembra comunque che neppure il Papa abbia potuto impedire che il suo organo inquirente accusasse Galilei di vehemente eresia e ordinasse di presentarsi per discolparsi davanti all'organo giudicante della Chiesa di Roma, al secolo il Santo Ufficio per discolparsi. Non avendolo fatto è stato condannato al carcere a vita,pena poi mitigata per la intercessione dei Medici, Sigg. di Firenze,mitigata con l'esilio perpetuo ove, come sapete, si spense (1642). Rispetto all'epilogo della vicenda di G.B. (1600) si avverte già una minore intransigenza . Ecco l'uomo che ha osato sfidare l'inquisizione:

Martin Lutero(1483-1546) teologo e riformatore

Galilei in dialogue on the Two Chief World Systems (1616), which parteciano three characters of the author. Sagredo (the same I know Galilei), Salviati (an architect friend of his), Simpli-cius (Pope Urban VIII) writes:

<Simplicius> I will confess that this whole night I went ruminating things yesterday and I really very beautiful new, and hefty considerations, despite all that I feel more stringer dall'oscu authority of so many great writers and in particular ..... You shake off Sig.Sagredo .... and sogghignate, ... as if to say some exorbitant.

<Sagredo> I just grin, but crediatemi that I burst in wanting to force it believes the laughter more for a fact that I grant.

<Salviati> Be good if there raccontiate to the fact that perhaps the Sig.Simplicio not continue to believe that having it moved by the laughter>

<Sagredo> There will never be safe, as we have simi-lar contradictory but this does not diminish Sig.Simplicius extravagance of peripatetic (Aristotle). Which is against sensible experience (referring to the Copernican Stat Sol) did not produce, then, is not pro-Dusse experience or other reasons, but the sole authority of the pure: Ipse dixit! Aristotle

<Simplicius> you have not purchased one large-de authority except for the strength of his shows-tions and for the depth of his speeches ....From this monologue on the Two Chief World Systems seems to conclude that Urban VIII had a lot of considerations to his interlocutor Galilei, while not accepting the same Stat Sol advocated. However, it seems that not even the Pope has potu-to prevent his vehement investigative body accused of heresy and ordered Galileo to appear to exonerate himself in front of the judging organ of the Church of Rome, aka the Holy Office to exonerate himself. Failing to do so has been sentenced to life imprisonment, then mitigated punishment for the intercession of the Medici, Sigg. Florence mitigation ta with perpetual exile where, as you know, he died (1642). Compared with the epilogue of the story of G.B. (1600) already feels less intransigence. Here is the man who dared to challenge the Inquisition:

Martin Luther(1483-1546)theologian and reformer

Martin Lutero(1483-1546) Si racconta che da gi giovane sotto la influenza della madre , donna pia ma incline alla superstizione che lo inviò alla scuola latina di Madgeburgo(1497),nel (1501) si laureo in filosofia. Dopo anni di insegnamento teologico in varie città. Predicando, con rigore morale secondo cui il cristiano si deve sempre riconoscere come un peccatore, giustificato nella penitenza.

Il suo carattere adamantino lo portò alla ribellione contro Roma Caput mundi della cristianità. La causa sta nel fatto che il domenicano Tetzel dava corso alla raccolta dei fondi per la costruzione di San Pietro in Roma in cambio della indulgenza plenaria. Un uomo del suo calibro non poteva non opporsi ad un tale mercato . Decise il 31 Ottobre del 1517 affisse sulle porte della cattedrale di Wittember 95 tesi redatte in latino e dirette contro quello che considerava una pratica contraria agli insegnamenti di Cristo con alle ricchezze materiali della Chiesa di Roma. Convenuto per discolparsi al legato pontificio, cardinale T.Caetano(1518). In una disputa con Jhonn Eck scrisse che '' anche il concilio ecumenisoco poteva errare e che il primato ponticio non era una istituzione divina, facendo infine appello al papa ''meglio informato''Il pontefice,Leone X invece, rimaso a lungo incerto, si decise,con la bolla Exsuge Domine(1520) a chiedere a Lutero una piena ritrattazione ,pena la scomunica Lutero rispose dando alle fiamme la bolla pontificia sulla piazzadi Wittemberg (1520). Venne perciò subito scomunicato che Lutero non tenne in alcun conto tanto che con un edito '' Alla nobiltà cristiana della nazione tedesca invita tutti gli uomi -ni per la salvezza della della Chiesa e della patria tedesca,atto formale della teologia luterana .

Secondo la quale lla salvezza dai peccati e solo opera di Dio.Nel 1521 convocato dalla dieta imperiale,Lutero si recò a Worms dove nonsoatante le pressioni si rifiutò di ritrattare. Nel 1534 compì la traduzione della Bibbia. In questa epoca insorge una disputa velenosa promossa da Erasmo da Rotterdam che lo accuso' di essere un terrorista per il" **libero arbitrio**" che andava insinuado nei cre-denti cristiani. La risposa ad Erasmo di Lutero arriva con un trattato ''**De servo encomio**'' Martin Lutero ha indotto la Chiesa(Concilio vaticano)da solo, di ritornare sulle orme di Cristo .

Martin Luther (1483-1546) is said to be already under the influence of the young mother, but prone to superstition pious woman who sent him to the school of the Latin Madgeburgo (1497) in (1501) he graduated in philosophy. After years of theological teaching in various cities. Preaching with moral rigor second-do which the Christian must always be recognized as a sinner, justified in penance. His character adamant led him to rebellion against Rome Caput mundi of Christianity. The cause lies in the fact that the Dominican Tetzel gave during the collection of funds for the construction of St. Peter in Rome in exchange for a plenary indulgence. A man of his caliber could not oppose such a market. He decided on October 31, 1517 posted on the doors of the cathedral of Wittenberg 95 theses written in Latin and directed against what he regarded as a practice contrary to the teachings of Christ with the material wealth of the Church of Rome. Agreed to exculpate the papal legate, Cardinal T.Caetano (1518). In a dispute with Jhonn Eck also wrote that'' the council ecumenisoco could err and that the primacy ponticio was not a divine institution, by finally appealed to the pope better informed'''' The bridge-fice, Leo X instead, left alone with long- uncertain, it de-cise, with the bubble Exsuge Domine (1520) to ask for a full retraction to Luther, on pain of excommunication Luther replied, setting fire to the papal bull on piazzadi Wittenberg (1520). It was therefore not immediately excommunicated Luther overrode so that with a published'' The Christian nobility of the German nation invites all men-tions for the salvation of the Church and the German homeland, a formal act of Lutheran theology. According to which lla salvation from sin and only by Dio.Nel 1521 convened by the Imperial Diet, Luther went to Worms where nonsoatante pressures refused to recant. In 1534 he made the translation of the Bible. In this era there arise a dispute poisonous promoted by Erasmus of Rotterdam who accuse him 'to be a terrorist for the "**free will**" insinuado that went into believers Christians. Remarries at The Erasmus Luther comes with a treatise'' **De servo praise** [a] Martin Luther led the Church (Second Vatican) alone, to return in the footsteps of Christ.

Spenta la luce della speculazione della grecia antica che simbolicamente facciamo coincidere con la uccisione di Archimede(212 a.C.) il letargo occidentale dura fino all'avvento dei padri fondatori a partire da **Copernico(Bacon-Galilei-Keplero-Newton)**.

L'europa all'alba del XV secolo è il teatro, Fig10-Sto1 in cui i poli scientifici foriscono nonostante l'ombra della inquisizio e lo spettro del Sato Ufficio che, come già ricordato, rappresenta la degenerazione del fanatismo.

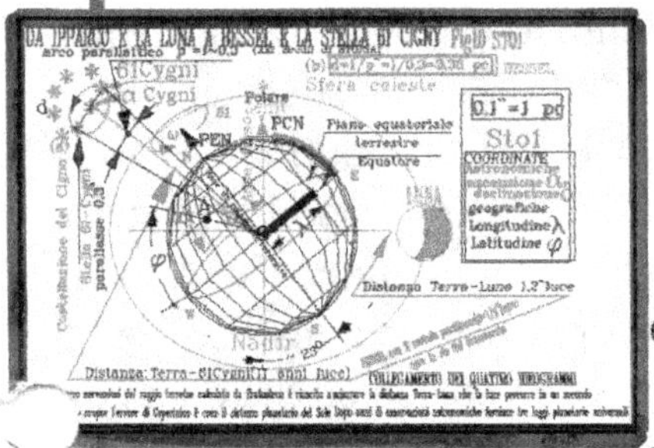

Usciamo ora da questo mare tempestoso .Per approdare al dominio del razionalismo scientifico che ha creato i presupposi di un dominio ampliado i confini del cosmo. Il primo ad uscire dal sistema solare è stato **Bessel**(1784-1846) che ha determinato la distanza della stella α-61Cigny dalla eclittica(piano della orbita terrestre), usando la sua formula definita dalla $d=\frac{1}{p''}$ (1) con d(distanza) e p'' parallasse della stella dimanica rispetto al sistema binario della sua dominante. L'arco di parallasse : p''=0,314'' da la $d=\frac{1}{0,314}=3,1847$ pc (parsec) che, tradotto in anni luce fornisce, di circa d=10.66 a.l

Per fissare le idee se si consideri l'orbita terrestre di Fig11-Tf1 , si può costatare che il raggio della orbita terrestre è di 100.000.000 di km che ,in metrica astronomica equivalgono ad un parsec(parallase secondo) Questa unità di misura fa riferimento ad un campione postoalla distanza di un parsec rispetto al quale visto da due opprtuna posizizioni dell'orbita teerestre vede appunto la parallasse di un sitema binario su si un piano parallelo alla ecclitica con un arco diameterale di un angolo p'' espresso in parsec . Tutti i cataloghi stellari riportano i sistemi binari applicando la (1) . Mappa Tav12-Fig13,costellazione della Ursa-Minoris

Off the light of speculation that the ancient greece to symbolic do coincide with the kill -ing of Archimedes (212 BC) Western hibernation lasts until the advent of the founding fathers from **Copernicus (Bacon - Galilei - Kepler - Newton**

. The dawn of the fifteenth century Europe was the theater, Fig.10 - STO1 , in which science parks flourish despite the shadow of inquisizio and the spectrum of Sato Office which , as already mentioned, is the degeneration of fanaticism. We go now from this stormy sea. Razionalis To arrive to the domain of mo - scientific considerations in mind which created a dominion ampliado the boundaries of the cosmos.

The first to leave the solar system was **Bessel** (1784-1846) which determined the distance of the star α 61–Cigny to the ecliptic (plane dell'0rbita terresstre) , using a formula defined by $d=\frac{1}{p''}$ (1 with d (distance) and p'' dimanica parallax of the star with respect to the binary system of its dominance. The arc of parallax : p = 0.314 '' from where $d=\frac{1}{0,314}=d = 3.1847$ pc (parsecs) which translated into light-years provides , with good approxithe distance in light years to d = 10.66 a.l

Fix ideas if consideriano the Earth's orbit of Fig11 - TF1 , one can notice that the radius of Earth's orbit is 100.000.000 km in metric astronomic equivalent to one parsec (parallase second) is unit of measurement is reference to a sample of a parsec postoalla distance with respect to which viewed from two giuste position dell'orbits terestrial sees the parallax of a gamin platform on a plane parallel to it eclitics diametra with an arc of an angle p'' expressed in parsec . A catalogs report the stellar binary systems by applying (1). Es.Tav12 - Fig13 , the constellation Ursa Minoris -

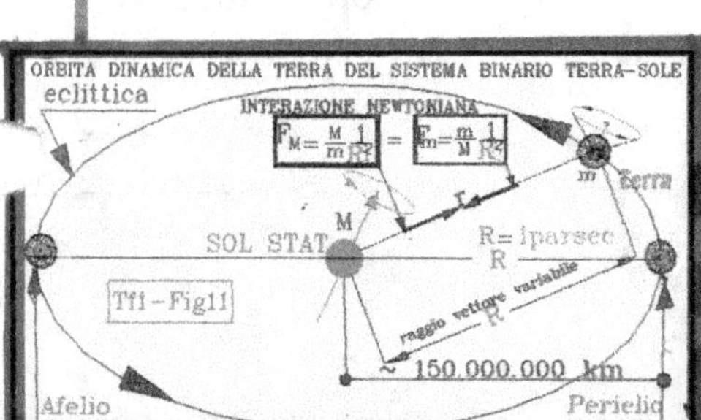

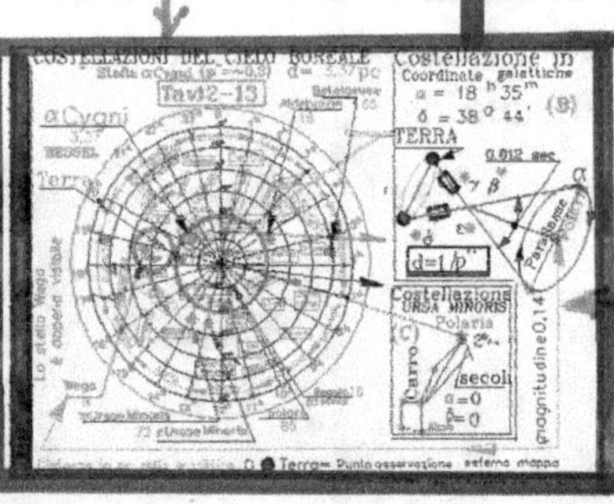

(La Terra il nostro pianeta)

[I] I movimenti della Terra e la forza di gravita'

I moti della TERRA,nicchia (b)**Fig3-Sto1** per quanto si riferisce alla descrizione a seguire,è considerata una sfera che descrive una ellisse con il **SOLE S** in un fuoco. La sua velocià scalare media è di ~ 29,79 kms^{-1}, il più importante , uno dei 14 moti del pianeta. Allo scopo della geodesia sono di interesse prevalente il moto di rotazione angolare di parallasse p'' asse Z,nicchia (b). La rotazione della Terra attorno al suo asse (polare) alla velocità angolare media in 464 ms^{-1}all'equatore, in quanto per la seconda legge delle aree di **Keplero** si ha: $\frac{dA}{dt}$=Cost. (1) come dire che lavelocità di rivoluzione dei pianeti è proporzionale alle aree descrtitte dal raggio vettore terrestre.Quanto al moto di precessione descritto dall'asse di rotazione della TERRA risulta di valore angolare p''=8",7946 da cui la distanza media TERRA-SOLE è di 149.598.500 km che si assume con valore medio di 150.000.000 miliondi di km e che gli astronomi assumono come unità di misura pari ad un parsec ed equivalente ad 3,25 anni luce :**d=1pc=3,25 a,l.** [2]

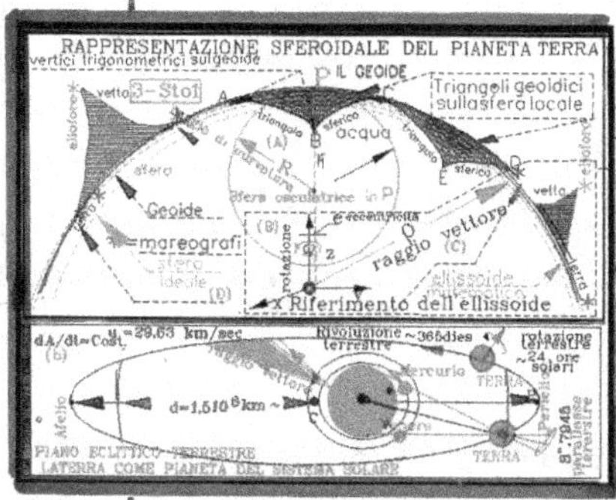

- **STRUTTURA RIGIDA E DEFORMABILE DELLA TERRA** -

La Terra considerata uno sferoide ,**Fig 3-Sto1**, ha una configurazione variabile dipendente dal rignonfia-mento ai poli. Per tale ragione non ha senso parlare di defformazione senza indicare l'algoritmo metrico cui far riferimento perciò premettiamo il calcolo di cui ci dovremo servire d'ora innanzi, Fig2 -Sto1. Questo quadro delle funzioni trigonometriche piane è la premessa per i calcoli dei triangoli e poligonali della rete topogragica che copre le terre di quattro continenti ma va esata estesa con le geodetiche(Linee tracciate sulla terrra supposta ellissoidica) . Con unaestrapola-zione in campo trigonometrico le stesse relazioni si applicano nel caso di triangoli sferici del piano di **Gauss** , fondatore dell'algebra complessa.Misurò la or-bita d Cerere ed inventò l'unità immaginaria j=$\sqrt{-1}$. determinando tutte le radici di una equazione di secon-do grado: $ax^2+bx+c=0$ [2]

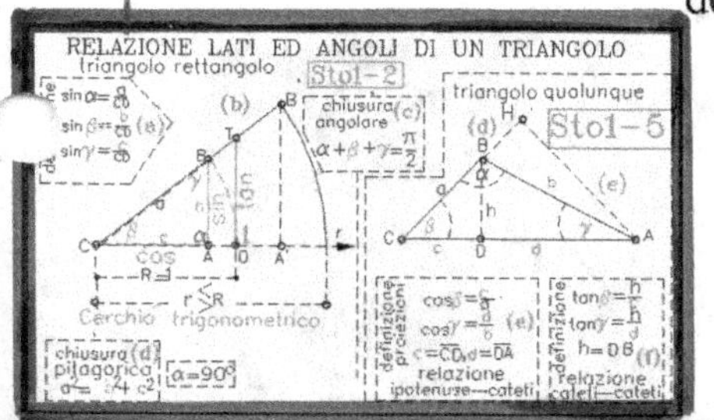

(The Earth our planet)

[I] Earth movements and strength of gravita'

I motions of the Earth , niche (b) **Fig3 - STO1** with regard to the following description , it is considered a sphere which describes an ellipse with the **SUN S** in a fire. His average is climbing the Speed ~ 29.79 kms^{-1}, the most important, one of the 14 motions of the planet. To the purpose of surveying are prevailing interest the motion of angular rotation of parallax p'' Z axis , niche (b). The rotation of the Earth about its axis (polar) to the average angular velocity in 464 ms-1all'equatore , as for the second law of areas **Kepler** has : $\frac{dA}{dt}$ = Const (1) as saying that lavelocità of revolution of the planets is proportional to the areas descrtitte terrestre.Quanto by the **radius vector** to precess described by the axis of rotation of the Earth is the angular value of p'' = 8 " , 7946 by which the distance mean Sun-Earth is 149,598,500 km which is assumed with mean value of 150,000,000 miliondi miles and that astronomers take as a unit of measurement equal to one parsec , equivalent to 3.25 years light : 1pc d = 1pc=3,26, a.l [2]

RIGID STRUCTURE OF THE EARTH AND DEFORMABLE -

The Earth regarded as a spheroid , **Fig 3 - Sto1** , STO1 , has a variable configuration dependent rignonfia -tion at the poles. for this reason, it makes no sense to talk about without defformazione indicate the algorithm metric which refer briefest therefore the calculation of which we will have to serve henceforth , **Fig2 - STO1** . This picture plane of the trigonometric functions is a prerequisite for the calculation of the triangles and polygonal network topogragica covering the lands of four continents but goes esata extended with the geode-sics (lines drawn on terrra supposed ellipsoid .) with an extrapolation in the same field trigonometric relationships apply in the case of spherical triangles of the Gaussian plane , founder of algebra complex . Again he measured the orbit of Ceres and invented the imaginary unit j = .$\sqrt{-1}$ determining all solution of a quadratic equation:

$$ax^2 + bx + c = 0 \quad [2]$$

[I] Il corpo ℜ dei numeri reali . Si indica con a,b,c... un numero scalare costante positivo o negativo . Ad es. a=2 o b=-2 che in valore assoluto (posivo) è definito di modulo : |-2|=+2. Quando si voglia indicare un numero variabile in valore al solito si indica con x , od altra come y.Se si vuole assumere la y come fuzione di x si scrive y=f(x) (1).

[II] Il corpo ℭ dei numeri complessi Per evitare di indicare che si tratta di un numero complesso costante si usa scrivere â mentre la variabile si indica al solito con z od s

[III] Alcune utili convenzioni del codice ASCI :
(Symbol : θ,ω,ε,...) # dec.139 ↔ dec.140 ← #
dec.142 → # dec.143 ↓ # dec.145 ± #
dec.133 ∞ # dec.119 ω # dec.178 ∅ #
dec.177 ≠ # dec.152 ÷ # dec.131 ≤ #

dec.165 ⊕ # dec.164 ≅ # dec.174 ∈
dec.175 ∉ # Ricordiamo che il significato algebrico sarà specificato al momento iniziale dell'uso.

[IV] Forme e proprietà dei vettori complessi. Si definisce forma rettangola del vettore o numero: $z=x+jy$ (1)

-La forma trigonometrica dei numeri complessi
Nella Fig12(a)Am11 è indicato un vettore del tipo esponenziale trigonometrico della anche cisoidale con evidente matrice lelettronica :come si può costatare la forma trigonometrica : $z=r[\cos(\phi)+j\sin(\phi)]$ (2), di componenti legate alla (1) dalla relazione : $r=\pm\sqrt{x^2+y^2}$. Infine la forma esponenziale: $z=\sqrt{x^2+y^2}\, e^{jz} =\sqrt{x^2+y^2}\,[\cos(\phi)+j\sin(\phi)]$ (3) , essendo e la base dei logaritmi naturale di **Nepero**(1550-1617), e=2,718281828 La (3) si generalizza per il teorema di **De Moivre** (1667-1744) .Questo consente di elevare z alla potenza n=1,2,..adoperadell'inglese A.De Moivre :
$z^n=\{r[\cos(\phi)+j\sin(\phi)]\}^n = r^n\,[\cos n\phi+j \sin n(\phi)]$ (4)

[I]RADICI DI NUMERI COMPLESSI Il numero w è la radice n-esima del numero z.La Fig1-Am11 rappresenta la : $w^n=z$ (5) .Allora : $w=z^{1:n}$ (6) .Dalla (3) si deduce:
$z^{1:n}=[\cos(\phi)+j\sin(\phi)]^{1:n}=r^{1:n}\{\cos\frac{\Phi+2k\pi}{n} + j \sin\frac{\Phi+2k\pi}{n}\}$ (7) k=1,2.....La (7) dimostra il teor. fond. per le radici della (6) , se $z\neq 0$

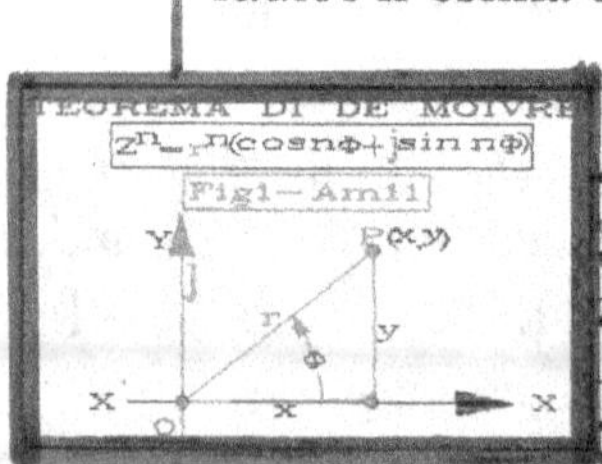

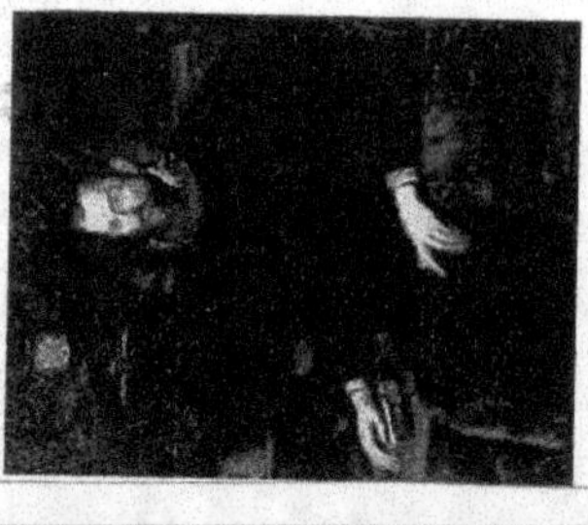
J. NEPERO

[I] The body ℜ of real numbers. Is denoted by a, b, c ... a scalar constant positive or negative. Eg . a = 2 b = -2 in absolute terms (posivo) is defined in the module: | -2 | = +2 . When you want to indicate a variable number in usual value is denoted by x , or other like y.Se you want to take the y as functio between x we write y = f (x) (1) . [II] The body C of complex numbers to avoid indicate that it is a complex number constant is used to write while the variable indicates usual with z or s

[III] Some useful conventions of the ASCI code

(Symbol : θ,ω,ε,...) # dec.139 ↔ dec.140 ← #
dec.142 → # dec.143 ↓ # dec.145 ± #
dec.133 ∞ # dec.119 ω # dec.178 ∅ #
dec.177 ≠ # dec.152 ÷ # dec.131 ≤ #

dec.165 ⊕ # dec.164 ≅ # dec.174 ∈
dec.175 ∉ # Recall that the one -fied algebraic signs will be specified at the time of the initial use.

[IV] Forms and properties of complex vectors.
It defines rectangular shape of the vector or number: $z = x + jy$ (1) - The trigonometric form of complex numbers in **Fig12 (a) AM11** is given a vector of exponential trigonometric also cisoidal with obvious matrix The electronics : how can ascertain the trigonometric form :
$z=r[\cos(\phi)+j\sin(\phi)]$ (2), components related to (1 by the relationship: $r =\pm \sqrt{x^2+y^2}$. Finally the exponential form :
$z = \sqrt{x^2+y^2}\, e^{Jz} = r\,[\cos (\phi)\ J\sin (\phi)]$ (3) , being and the base of the natural logarithms of **Napier** (1550-1617) , e = 2.718281828 . The equation (3) is generalized to the heorem of **De Moivre** (1667-1744. This admit to raise z to the power n=1,2.....,
$z^n=\{r[\cos(\phi)+j\sin(\phi)]\}^n = r^n\,[\cos n\phi+j \sin n(\phi)]$ (4

[I] ROOTS OF COMPLEX NUMBERS
The number w is the nth root of the number z. The Fig1 - AM11 represent : $w^n = z$. (5)
Then: $w = z^{1:n}$ (6) From (3) it follows :
$z^{1:n}=[\cos(\phi)+j\sin(\phi)]^{1:n}=r^{1:n}\{\cos\frac{\Phi+2k\pi}{n} + + j \sin\frac{\Phi+2k\pi}{n}\}$ (7

k = 1,2 ..... the (7) of the theor - show . fond . for the roots of (6) , if $z\neq 0$

DE MOIVRE
(1667 - 1744)

[I]Se ad ogni valore z della variabile complessa corrisponde uno o più valori della funzione w si indica $w=f(z)$ (1) si ottiene la funzione f di variabile z. Es. se si pone $z=-1$ si trova : $w=(-1)^2=+1$, $w=\sqrt{-1}=j$ (1_2) . La (1) sarà nel seguito intesa ad un sol valore

[II]Funzioni inverse. Dalla w=f(z) si può considerare la funzione $z=g(w)=f^{-1}(w)$ (2),inversa della (1) .Mentre nella (1) la variabile $z\in\mathfrak{D}$ nella (2) $w\in\mathfrak{D}$ di g(w).

[III] TRASFORMAZIONI Se: $w=u+jv$ (3)con u,v reali allora la (3) è una funzione ad un sol valore della vaiabile complessa : $z=x+jy$ è lecito scrivere la equivalenza: $u+jv=f(x+jy)=w$ (4).Allora eguagliando fra loro le parti reali ed immaginarie si ottiene la corrispondenza : $u=u(x,y)$ (5) e $v=v(x,y)$ (6) . La

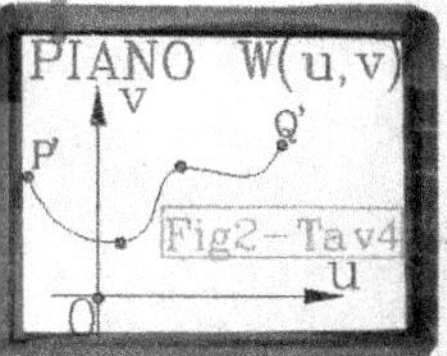

Fig2-Tav4, rappresenta il piano**W**(di coordinate u,v)dei punti $\{P', Q', ..\}\in W(u,v)$ Le equazioni (5) e (6) sono le trasformate di $W(u,v)=f(z)$ (7)rappresentata dalla (4)

[III] COORDINATE CURVILINEE

Data la trasformazione $W(u,v)=f(z)$ o le equivalenti (5),(6) si dice che $P(x,y)\in W(u,v)$ sono le coordinate ortogonali x,y delle curvilininee **(u,v)** , Fig13-Tav4.

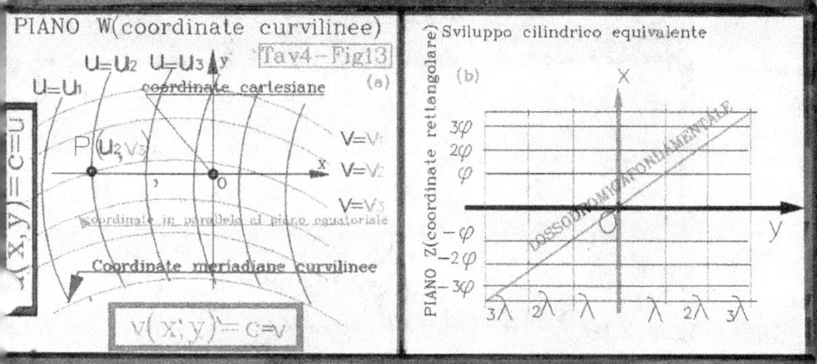

Le linee $u(x,y)=c_1$,$v(x,y)=c_2$, con c_1 e c_2 costanti sono definite corrdinate ed ogni coppia di queste linee si intersecano in un punto. Queste linee del piano W si trasformano,come vederemo, in rette mutuamente ortogonali come dimostra la figura della nicchia **(b)**

[IV]Le funzioni elementari dello spazio $\mathcal{C}$

Dato il polinomio: $w=a_0z^n+a_1z^{n-1}+....+a_{n-1}z+a_n$ (8) con i fattori: $\{a_0, a_1, ..., a_{n-1}, a_n\}$ (8') costituiti da numeri complessi ed n un intero naturale detto grado del polinomio P(z). La trasformazione $w=az+b$ (9) si dà il nome di lineare

[I] If, for each value of the complex variable z corresponds to one or more values of the function w -cates $w=f(z)$ (1), we obtain the function f of the variable z . Eg if you put $z=-1$ is : $w=(-1)^2=1$, $w=\sqrt{-1}=j$ (1_2) . Equation (1) is intended to result in a single value

[II] Inverse functions . By w= f (z) we can consider the function $z=g(w)=f^{-1}(w)$ (2) , inverse knows of (1) . While in (1) the variable $z\in\mathfrak{D}$ (2) $w\in\mathfrak{D}$ of g (w) .

[III] TRANSFORMATIONS If $w=u\,jv$ (3) with u, v real then (3) is a function to a single value of vaiabile complex $z=x\,jy$ is permissible to write the equivalence : $u+jv=f(x+jy)=w$ (4) . Then equaling between them the real and imaginary parts is obtained correspondence: $u=u(x,y)$ (5) and $v=v(x,y)$ (6) . The Fig2 - Tav4 , represents the pianoW (coordinates u , v) of the points { **P '**, **Q '** , ..} The equations (5) and (6) are the transforms of $W(u,v)=f(z)$ (7) represented by (4)

[III] GIVEN THE TRASFORMAZIONEW CURVILINEAR coordinates $W(u,v)=f(z)$ or the equivalent (5) , (6) it is said that $P(x,y)\in W(u,v)$ are the orthogonal coordinates x , y of curvilininee (**u** , **v**) , Fig13 - Tav4 .

The lines $u(x,y)=c_1$, $v(x,y)=c_2$, with c1 and c2 are constants defined corrdinate and each pair of these lines intersect in a point . These lines of the plane W are transformed, as we will see , in mutually orthogonal straight lines as shown in the figure of the opening **(b)**

[IV] The elementary functions of the space $\mathcal{C}$

Since the polynomial $w=a_0z^n+a_1z^{n-1}+....+a_nz+a_n$ (8) with the factors : $\{a_0, a_1, ..., a_{n-1}, a_n\}$ (8 ') consisting of complex numbers and n a natural number called the degree of the polynomial P (z). The transformation $w=az+b$ (9) gives the name of linear

[I]Funzione rapporto fra polinomi.

Dati i polinomi lineari: $P(z)=az+b$ (1), $Q(z)=cz+d$ (2) Si dice trasformazione razionale : $w=P(z)/Q(z)$ (3) il raporto,cioè: $w=\frac{az+b}{cz+d}$ (4) con ad-bc≠0 è per definizione una trasformazione frazionaria lineare.

[II]Funzioni esponenziali. Questo inportante spazio algebrico è definito dalla funzione:

$w=e^z=e^{x+jy}=e^x.e^{jy}=e^x(\cos y+j\sin y)$ (5) .Questa funzione w è ottenuta con la base e=2,718281828>0 dei logaritmi neperiani,elevata allapotenza z=x+jy e, dato che il prodotto di due potenze della stessa base è uguale ad una potenza conla stessa base ed esponente la somma degli esponenti risulta giusificata la eguaglianza $e^{x+jy}=e^x.e^{jy}$. Ricordando la trasformata di **De Moivre** $e^{jy}=(\cos y+j\sin y)$ (5_1),con y un parametro angolare.Anche: $w=a^z=a^x.a^{jy}=a^x(\cos y+j\sin y)$ (6) se a>0.Nel caso a=10 la base è dei logaritmi decimali In generale- I numeri z1 e z2 esponenziati godono delle stesse prprietà dei nuneri reali,cioè :

$e^{z1}.e^{z2}=e^{z1+z2}$ (6_1), $e^{z1}/e^{z2}=e^{z1-z2}$ (6_2)

[III]Funzioni trigonometriche in formato Expo.

$\sin z=\frac{e^{jz}-e^{-jz}}{2j}$ [1], $\cos z=\frac{e^{jz}+e^{-jz}}{2}$ [2]

$\sec z=\frac{1}{\cos z}=\frac{2}{e^{jz}+e^{-jz}}$ [3], $\csc z=\frac{1}{\sin z}=\frac{2j}{e^{jz}-e^{-jz}}$ [4]

$\tan z=\frac{\sin z}{\cos z}\frac{e^{jz}-e^{-jz}}{j(e^{jz}+e^{-jz})}$ [5] $\cot z=\frac{\cos z}{\sin z}\frac{j(e^{jz}+e^{-jz})}{e^{jz}-e^{-jz}}$ [6]

[IV]Proprietà delle funzioni trigonometriche (sono formalmente dedotte dalle reali)

$\sin^2 z+\cos^2 z=1$ (7) , $1+tn^2 z=\sec^2 z$ (8) , $1+\cot^2 z=\csc^2 z$ (9)

$\sin(-z)=-\sin z$ (10) , $\cos(-z)=\cos(z)$ (11), $\tan -z=-\tan z$ (12)

$\sin(z_1\pm z_2)=\sin z_1\cos z_2\pm\cos z_1\sin z_2$ (13)

$\cos(z_1\pm z_2)=\cos z_1\cos z_2\ \ \sin z_1\sin z_2$ (14)

$an(z_1\pm z_2)=\frac{\tan z_1\pm\tan z_2}{1+\tan z_1\tan z_2}$ (15)

[V]Le equivalenze trignometriche↔iperboliche

$\sin jz=j\sinh z$ (16), $\cos jz=\cosh z$ (17),

$\tan jz=j\tan hz$ (18) → $\sinh jz=j\sin z$ (19),

$\cosh jz=\cos z$ (20), $\tanh jz=j\tan z$ (21)

[I] Function ratio of polynomials.

Given the linear polynomials : $P(z)=az+b$ (1) , $Q(z)=cz+d$ (2) It is said rational transformation : $w=P(z)/Q(z)$ (3) report it , ie : $w=\frac{az+b}{cz+d}$ (4) with ad- bc≠ 0 is therefore by definition a linear fraction transformation **[II] exponential Functions** .

This inportante algebraic space is defined by the function: $w=e^z=e^{x+jy}=e^x.e^{jy}=e^x(\cos y+j\sin y)$ (5 This function w is obtained with the base e = 2.718281828 > 0 of logarithms neperiani , high allapotenza z = x jy and , since the product of two powers of the same base is equal to a power thelates same base and exponent the sum of the exponents is giusificata the identity $e^{x+jy}=e^x.e^{jy}$. . Recalling the transform of **De Moivre** $e^{jy}=(\cos y+j\sin y)$ (5_1) with a parameter y angolar and:

$w=a^z=a^x.a^{jy}=a^x(\cos y+j\sin y)$ (6) if a> 0. If a = 10 is the base of logarithms decimal . In general the numbers z_1 and z_2 esponenziati enjoy the same prprietà of nuneri real , that is :

$e^{z1}.e^{z2}=e^{z1+z2}$ (6_1) $e^{z1}/e^{z2}=e^{z1-z2}$ (6_2)

[III]Function exponential

$\sin z=\frac{e^{jz}-e^{-jz}}{2j}$ [1], $\cos z=\frac{e^{jz}+e^{-jz}}{2}$ [2]

$\sec z=\frac{1}{\cos z}=\frac{2}{e^{jz}+e^{-jz}}$ [3], $\csc z=\frac{1}{\sin z}=\frac{2j}{e^{jz}-e^{-jz}}$ [4]

$\tan z=\frac{\sin z}{\cos z}\frac{e^{jz}-e^{-jz}}{j(e^{jz}+e^{-jz})}$ [5] $\cot z=\frac{\cos z}{\sin z}\frac{j(e^{jz}+e^{-jz})}{e^{jz}-e^{-jz}}$ [6]

[IV] Properties of trigonometric functions

(they are formally deduced from the real)

$\sin^2 z\ \cos^2 z=1$ (7), $1=tn^2 z\ \sec^2 z$ (8), $1+\cot^2 z\ \csc^2 z$ (9)

$\sin(-z)=-\sin z$ (10), $\cos(-z)=\cos(z)$ (11), $\tan -z=-\tan z$ (12)

$\sin(z_1\pm z_2)=\sin z_1\cos z_2\pm\cos z_1\sin z_2$ (13)

$\cos(z_1\pm z_2)=\cos z_1\cos z_2\ \ \sin z_1\sin z_2$ (14)

$an(z_1\pm z_2)=\frac{\tan z_1\pm\tan z_2}{1+\tan z_1\tan z_2}$ (15)

[V] The equivalences trignometric ↔hiperbolic

$\sin jz=j\sinh z$ (16), $\cos jz=\cosh z$ (17),

$\tan jz=j\tanh z$ (18) → $\sinh jz=j\sin z$ (19)

$\cosh jz=\cos z$ (20), $\tanh jz=j\tan z$ (21)

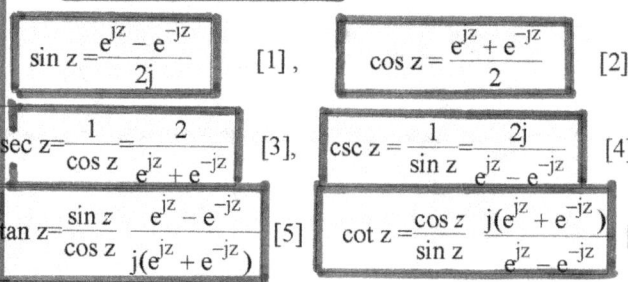

Un vettore che sin dai tempi più antichi si proponenva al pensiero degli studiosi era il moto dei corpi nello spazio fisico più o meno rapidi detto velocià vettoriale $\bar{u}$, mentre u (cioè il suo modulo o gradezza) è detta velocità scalare.La Fig10-Am24 mostra il razzo vettore

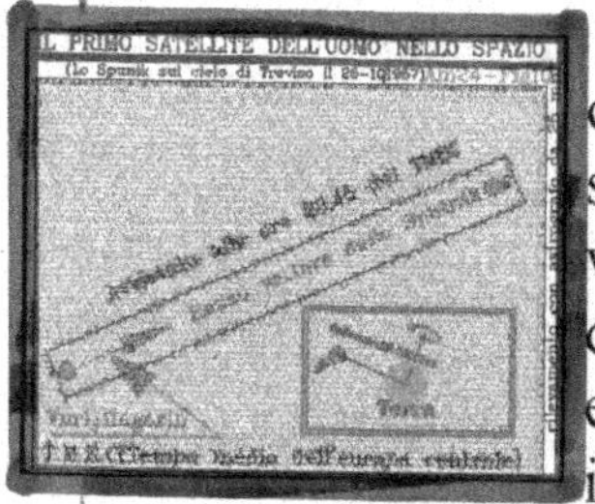

che ha aperto l'era dei voli nello spazio sede dei corpi celesti. La velocità della navicella Soyuz , vista da terra come un asteroide , oltre sd essere un vettore,in tal senso $\bar{u}$ varia, in dipendenza della direzione e del verso $\vec{u}$ o $\overleftarrow{u}$. Se la intensità scalare è costante, cioè : $|\vec{u}|$=Cost. il moto della Soyuz è circolare uniforme .Se $|\vec{u}|\neq$Cost. allora la traiettoria è ellittica.Di proposito siamo partiti da un caso significativo per l'uomo faber capace di dar corpo alla tecnologia meccanica dei satelliti (oggi a migliaia solcano il cielo del nostro pianeta e non sempre a fini scientifici.Quanto dettoper il vettore velocità vale anche per altri vettori fisici.

[I] La struttura dei campi vettoriali e scalari

Indichiamo le componenti cartesiane vettoriali nella forma : {Φx,Φy,Φz} (a) e tenendo presente che i vettori reali(velocità, forza,momento,ect,ect) sono di natura intrinseca o fisica indipnendenti dal sistema con il quale si rappresentano e si misurano.

[II] Rappresentazione di campo scalare. Si assuma a priori due sistemi di coordinate{x,y,z} (b){x',y',z'}(c)

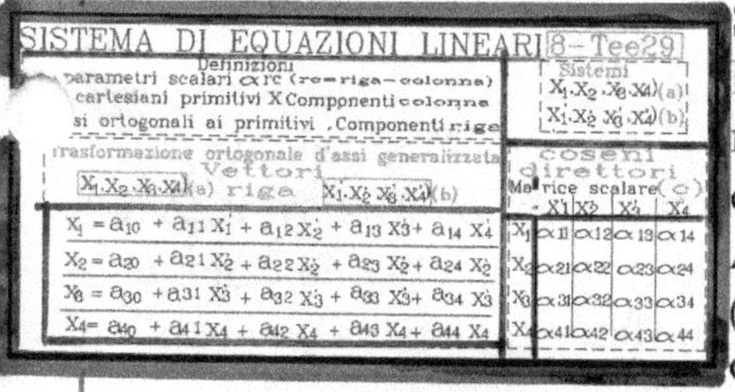

Si ottiene il campo Fig8-Tee29 rappresenta il sistema(b) per mezzo dei coseni direttori (c) Allora per mezzo delle (a),cioè le :{ Φx(x,y,z), Φy(x,y,z),Φz(x,y,z)}(d)

[III]La trasformazione di un campo vettoriale

$$\begin{aligned} \Phi x &= a_{11}\Phi x' + a_{12}\Phi y' + a_{13}\Phi z' \\ \Phi y &= a_{21}\Phi x' + a_{22}\Phi y' + a_{23}\Phi z' \\ \Phi z &= a_{31}\Phi x' + a_{32}\Phi y' + a_{33}\Phi z' \end{aligned} \quad [A]$$

Con analogo procedimento la trasformazione del campo vettoriale delle componenti note (d)

$$\begin{aligned} \Phi x' &= a_{11}\Phi x + a_{12}\Phi y + a_{13}\Phi z \\ \Phi y' &= a_{21}\Phi x + a_{22}\Phi y + a_{23}\Phi z \\ \Phi z' &= a_{31}\Phi x + a_{32}\Phi y + a_{33}\Phi z \end{aligned} \quad [B]$$

A vector that since ancient times is proponenva at the thought of the scholars was the motion of bodies in physical space more or less rapid said the Speed vector , while u (ie its desired size or form) is said to speed scalare.La Fig10 - AM24 shows the carrier rocket which opened the era of space flight headquarters of celestial bodies. The speed of the Soyuz spacecraft , seen from earth like an asteroid , as well as sd be a vector, in this sense varies $\bar{u}$, depending on the direction $\vec{u}$ and $\overleftarrow{u}$ toward or. If the intensity scale is constant, ie $|\vec{u}|$=Cost. the uniform circular motion of the Soyuz is If $|\neq$Cost. | well then the trajectory is about ellitivall.Di we started from a significant case for man faber able to give shape to the mechanical technology of satellites (now in thousands across the sky of our planet and not always for the purpose scientifici.Quanto dettoper the velocity vector also applies to other physical vectors .

[I] the structure of the vector and scalar fields

Denote the Cartesian components of vector in the form: { Φx, Φy , Φz } (a) and bearing in mind that the real vectors (speed, force, momentum, etc., ect) are of intrinsic nature or physical indipnendenti from the system with which they represent , and measure .

[II] Representation scalar field . Assume a priori two systems of coordinates{x,y,z}(b),{x', y', z'}(c) is obtained the field fig8 - Tee29 simbolyze the system (b) by means of the direction cosines (c) Then by means of (a), namely: { Φ_x (x,y,z), Φ_y(x, y,z) , Φ_z (x,y,z) } (d)

[III] The transformation of a vector field

$$\begin{aligned} \Phi x &= a_{11}\Phi x' + a_{12}\Phi y' + a_{13}\Phi z' \\ \Phi y &= a_{21}\Phi x' + a_{22}\Phi y' + a_{23}\Phi z' \\ \Phi z &= a_{31}\Phi x' + a_{32}\Phi y' + a_{33}\Phi z' \end{aligned} \quad [A]$$

With a similar procedure, the transformation of the components of the vector field notes (d)

$$\begin{aligned} \Phi x' &= a_{11}\Phi x + a_{12}\Phi y + a_{13}\Phi z \\ \Phi y' &= a_{21}\Phi x + a_{22}\Phi y + a_{23}\Phi z \\ \Phi z' &= a_{31}\Phi x + a_{32}\Phi y + a_{33}\Phi z \end{aligned} \quad [B]$$

Data una funzione f(x,y,z)data rispetto ad u sistema di assi c.o. Il campo vettoriale che ha per componenti le derivate parziali prime indefinitamente derivabili , di componenti $\{ \text{grad}_P\, f(x,y,z) \equiv (\frac{\partial f}{\partial x}, \frac{\partial f}{\partial y}, \frac{\partial f}{\partial z})\}$ (1) esprime il gradiente del campo scalare nei punti della regione $\mathcal{R}$ in cui risulta verificata la (1).Una notevole proprietà del gradiente f_P (x,y,z) è che se P si sposta di dP esso ammette : $df = \text{grad } f \times dP = \frac{\partial f}{\partial x} dx + \frac{\partial f}{\partial y} dy + \frac{\partial f}{\partial z}$ (2) per differe nziale se poi inchiamo con $\underline{n}$ il versore di dP=n dn la (1) può esprimersi nella forma $\frac{df}{dn} = \text{grad } f \times dn$ (3) Le proprietà del gradiente sono implicite nella(1) in ogni punto $P \in \mathcal{R}$. Infatti per f=Cost. grad f=0 (3_1) , (f_1+f_2)se$\in \mathcal{R}$ si ha : grad (f+g)= grad f + grad g (3_2) grad (f·g)=f gradg+g gradf (3_3) , grad $F(r) = \frac{dF}{dr}$ grad r (3_4) Proprietà differenziali della funzione gradienete che ha le stesse proprietà delle derivate ordinarie con la sola differenza che in questa sede sono adattate alla soluzione di propblemi fisici, in particolare al modo dei mezzi continui Ad es. nella , **Fig11-Am11**, è rappresentato il fluss di un vettore di un campo

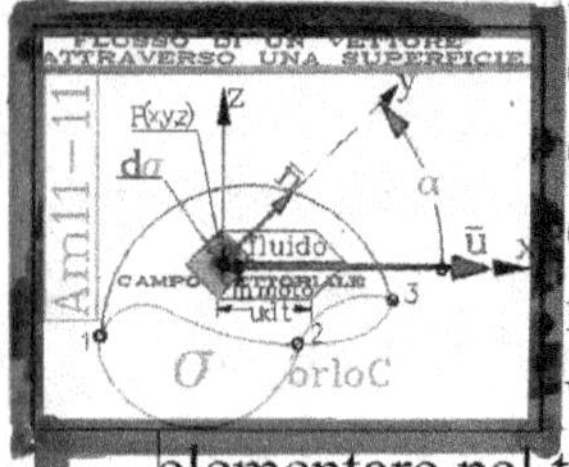

fluido ma che può .come vedremo estendersi concettulamanate ai campi elettroma- gnetici di onde di pura energia senza il trasporto di particelle materiali. Il prisma **d**σ supposto di velocità $\bar{u}$ descriverà il volume elementare nel tempo dt per cui sarà:

$d\sigma \cdot u\, dt \cos\alpha = = u_n\, d\sigma\, dt$ (4) Per una regione $\mathcal{R}$, in ogni punto P del quale risulta soddisfatta la(1) della superficie di frontiera $\mathbb{S}$ è possibile integrare il flussso del vettore $\bar{u}$ nel modo : $\int_\sigma \rho\, u_n d\sigma$ (5)

[I] I CAMPI DI C.F. GAUSS (1778-1855) Nelle precedenti pg 47 e 48 relative al corpo dei numeri complessi mostrano il prodigio della unità $j=\sqrt{-1}$ (a) immaginaria che, dal corpo dei numeri reali si è estesa all'universo dei numeri complessi ,aprrendo le porte all'algoritmo che ha dato alla fisica matematica la sua veste attuale dato che la **(a)** rappresenta la quarta dimensione che unisce lo spazio al tempo(**Einstein**).Lemma di **Gauss** della trasforma-zione di integrali**Data la funzione f(x,y.z)**.

Se f(x,y,z) è continua con le sue derivate prime nello spazio S si trovano le equazioni:

$$\int_S \frac{\partial f}{\partial x} dS = \int_\sigma f\, \alpha\, d\sigma \quad (61), \quad \int_S \frac{\partial f}{\partial y} dS = \int_\sigma f\, \beta\, d\sigma \quad (62), \quad \int_S \frac{\partial f}{\partial z} dS = \int_\sigma f\, \gamma\, d\sigma \quad (63)$$

Given a function f (x , y, z) given with respect to u axis system co The vector field which has components for the first partial derivatives indefinitely derivable , components $\{ \text{grad}_P\, f(x,y,z) \equiv (\frac{\partial f}{\partial x}, \frac{\partial f}{\partial y}, \frac{\partial f}{\partial z})\}$ (1) expresses the gradient of the scalar field at points in the region R in which it is verified (1). a remarkable property of the gradient f_P (x,y,z)is that if P moves dP it admits $df = \text{grad } f \times dP = \frac{\partial f}{\partial x} dx + \frac{\partial f}{\partial y} dy + \frac{\partial f}{\partial z}$ (2) to differen tial if we ito show with the unit vector of the dP =n dn(1) can be expressed in the form $\frac{df}{dn} = \text{grad } f \times dn$ (3) The properties of the gradient are implicit in (1) at any point $P \in \mathcal{R}$.In fact, for f =Const is grad f= 0(3_1) (f_1+f_2)if$\in \mathcal{R}$ we have grad (f+g)=grad f +grad g (3_2) grad (f+g)=f grad g+g grad f (3_3),$F(r) = \frac{dF}{dr}$grad r (3_4) Property gradienete differential of the function that has the same properties of ordinary derivatives with the only difference that in this seat are adapted to the solution of propblemi physical , in particular to the way of continuous media eg . in , **Fig11 - AM11** , shows the Flux of a vector field of a fluid , but he can. concettulamanate as we shall extend to the electromagnetic fields of waves of pure energy without the transport of material particles.The prism **d**σ but supposed to speed $\bar{u}$ describes the elementary volume in the time interval dt, so it will be : $d\sigma \cdot \bar{u}\, dt \cos\alpha = u_n\, d\sigma\, dt$ (4)

For a region R , in each point P of which is met (1) of the surface boundary $\mathbb{S}$ is possible to integrate the carrier flussso as : $\int_\sigma \rho\, u_n d\sigma$ (5)

[I] FIELDS CF Gauss (1778-1855) In previous, pg 47 and 48 , concerning to the body of complex numbers show the miracle of unity imaginary, $j=\sqrt{-1}$ (a) from the body of the real numbers has expanded the universe of complex numbers, the doors to aprrendo ' algorithm to mathematical physics which gave its present form since the (a) represents the fourth dimension that combines the space -time (Einstein) Lemma of Gauss transformation of integrals of **the function f (x,y,z).**

If f(x,y,z) is continuous with its first derivate in space S are the equations :

$$\int_S \frac{\partial f}{\partial x} dS = \int_\sigma f\, \alpha\, d\sigma \quad (61), \quad \int_S \frac{\partial f}{\partial y} dS = \int_\sigma f\, \beta\, d\sigma \quad (62), \quad \int_S \frac{\partial f}{\partial z} dS = \int_\sigma f\, \gamma\, d\sigma \quad (63)$$

Riprendiamo quanto lasciato in sospeso nel paragrafo VI di pg-46(esponenziali complesse).

[I] Funzioni esponenziali particolari
La funzione z^{α} dove α può essere complesso è definita dalla $e^{\alpha \ln z}$ (1). In campo reale sussiste la equivalenza $y=e^x \leftrightarrow x=e^{\ln y}$ (2), se y>0. In analogo modo se f(z) e g(z) sono funzioni di z date si può definire la identità : $f(z)^{g(z)}=e^{g(z)\ln f(z)}$ (3) se f(z)>0. **[II]Funzioni algebriche e funzioni trascendenti .**Se w è una soluzione della equazione algebrica $Po(z)w^n+(z)w^{n-1}+...+P_{n-1(z)}w+Po(z)=0$ (4) (4) si dice che w=f(z) è una funzione algebrica di z. Nella {Po≠0 ,$P_1(z)$,...,Pn(z) } sono polinomi in z ed n un intero positivo. Es. Sia $w=z^{1/2}$ una soluzione della equazione : $w^2-z=0$ (4_1). Infatti $w=\sqrt{z}$ che verifica la (3_1).Ogni funzione che non possa essere espressa da una soluzione tipo(3) è detta trascendente Le funzioni logaritmiche,trigonometriche,iperboliche di pg-48 e le loro inverse sono funzioni trascendenti elementari

[III] PUNTI E RETTE DI DIRAMAZIONE
Supponiamo di avere la funzione : $w=z^{1/2}$ (4) Se il punto A percorre un giro completo in senso antiorario partendo da A si ottiene la funzione $z=re^{j\theta}$ (5), Fig2-Tfc1 . La (4) si trasforma in: $w=\sqrt{r}\,e^{j(\theta/2)}$ (6). Quindi si ha: A,θ=θ1 è $w=\sqrt{r}\,e^{j(\theta 1/2)}$ Quando,dopo un giro completo si ritorna in A.Per la: $\theta=\theta_1+2\pi$(a) $w=\sqrt{r}\,e^{(2\pi)/2}=-\sqrt{r}e^{j\theta 1/2}$ (7) con un valore diverso da quello di partenza.Se si compie un altro giro completo fino al punto A si ha $\theta=\theta_1+4\pi$ e quindi : $w=\sqrt{r}\,e^{j(\theta 1+4\pi)/2}=-\sqrt{r}\,e^{j\theta 1/2}$ (8) identico alla (7).

[III]SUPERFICI CONTINUE DI B.Riemann (1826-1866) . Se f(z) è continua in una regione ℜ. Allora per ogni $z_o \in ℜ$ esiste un numero ε>0 tale che il numero δ>0 verifichi la condizione : $|z-zo|<\delta$ cui consegue la relazione $|f(z)-f(zo)|<\varepsilon$ (9) . Se è possibile determinare una d funzione di e,manon del particolare punto z_o, allora si dice che f(z) è uniformemente continua nella regione si ℜ se f(z)è uniformemente continua in ℜ per ogni ε>0 èpossibile determinare una δ>0 tale che $|z_1-z_2|<\delta$ verificala la $|f(z_1)-f(z_2)|<\delta$ (10) Esercizio Fig21.Se c_1 e c_2 sono costanti reali del piano z .Sono rappresentati i punti di rette $u=c_1$,$v=c_2$,piano $w=z^2$ (11) per:{c1=2-4,-2,-4},{c2=2=2,4,-2,-4}sono iperboli asitotiche equilatere,assi x,jy

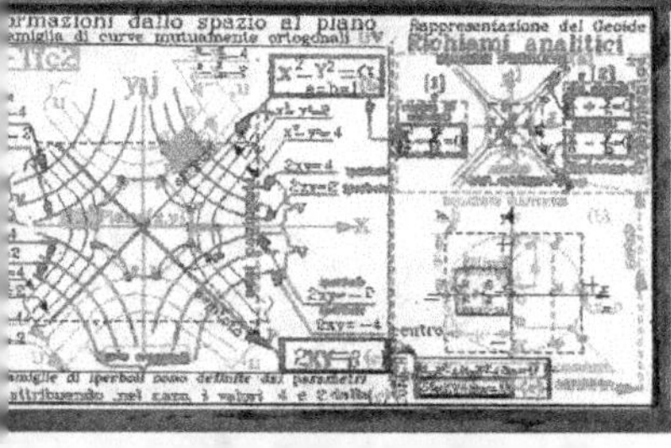

GAUSS AND SPACE COMPLEX pg -51

We take what was left suspended in paragraph VI of pg -46 (complex exponential functions) .

[I] Exponentials The z^{α} where a particular function can be complex is defined by : $e^{\alpha \ln z}$ (1). In the field there is a real equivalence $y=e^x \leftrightarrow x=e^{\ln y}$ (2) if y > 0 . In similar manner,if f(z) and g(z) are functions of z dates can define the identity : $f(z)^{g(z)}=e^{g(z)\ln f(z)}$ (2) where f (z) > 0 .

[II]algebraic functions and transcendental functions
If wis a solution of algebraic polynomial equation : $Po(z)w^n+P_1(z)w^{n-1}+....+P_{n-1(z)}w+Po(z)=0$ (3) it is said that w=f (z) is an algebraic function of z . In (3) {Po≠0 ,$P_1(z)$,...,Pn(z) } are polynomials in z and n a positive integer. Eg Let $w=z^{1/2}$ a solution of the equation : $w^2-z=0$ (3_1). In fact $w=\sqrt{z}$ that verification (3_1) . Each function that is imposible be expressed as the solution of (3) is called transcendental . Logarithmic functions , trigonometric , hyperbolic of pg -48 and their inverses are elementary transcendental functions

[III] POINTS AND LINES OF BRANCH
Suppose we have the function: $w=z^{1/2}$ (4) If point A along a full turn counterclockwise starting from A function is obtained: $z=re^{j\theta}$ (5) . Fig2 - Tfc1.La (4) is transformed into $w=\sqrt{r}\,e^{j(\theta/2)}$ (6) Then we have: A, θ = θ1 is $w=\sqrt{r}\,e^{j(\theta 1/2)}$. When , after a complete swing is back in Aing it to : $\theta=\theta_1+2\pi$ (a) $w=\sqrt{r}\,e^{j(\theta 1+2\pi)/2}=-\sqrt{r}\,e^{j\theta 1/2}$ (7)with a different value of daquello partenza.Se takes another full circle to the point a we have θ= θ1+4 p and then : $w=\sqrt{r}\,e^{j(\theta 1+4\pi)/2}=-\sqrt{r}\,e^{j\theta 1/2}$ (8) identical at the previous value .

[III] SURFACESOFCONTINUOUSB.RIEMANN
If f (z) is continuous in a region ℜ. Then for every $z_o \in ℜ$ exists a number ε> 0 such that the number δ > 0 verifies the condition : $|z-z_o|<\delta$ which follows the relation $|f(z)-f(zo)|<\varepsilon$ (9). If you can determine a function of d and , manon of the particular point z_o , then we say that f(z) is uni- formaly continuous in the region P . Conversely , if f (z) is uniformly continuous in ℜ for each ε> 0 you can determine a δ>0such that $|z_1-z_2|<\delta$ verifies the $|f(z_1)-f(z_2)|<\delta$ (10) (10) Esercizio.Fig21- Tfc2. Siano real constants c1 and c2 of the z-plane . are represented the points of the straight lines u=c1 and v = c2 = c1 of the plane $w=z^2$ (11) : for: c1 = { 2-4, -2, -4 } and { c2 =2,4, -2, -4 } is equilateral hyperbolas asymptotic to the axes x , jy

[I] Le equazioni algebriche di secondo grado :
Poniamo $ax^2+bx+c=a(x-x_1)(x-x_2)=0$ (1) .Il primo membro, per il **T.F.A.**ammette sempre due radici che la soddisfano identicamente , cioè x_1 ed x_2. Il problema è proprio la determinazione dei valori delle radici od autovalori.Nella(1)si identifica il numero $\Delta=b^2-4ac$ (2) detto **discriminante** del primo membro della (1) equivalente al secondo membro fattorizzata In generale gli autovalori, secondo teoria risultano :

$$x_{1,2}=\frac{-b\pm\sqrt{b^2-4ac}}{2a}$$

(3) . Le (3) sono state risolte co- me equazione algebrica di terzo grado da Tartaglia, **Niccolò Fontana**(1499-1557)da cui è deducibile la (2). Le soluzioni si diversificano con il valore del Δ. **Primo caso** : Sia $\Delta=b^2-4ac=0$,allora se i parametri a,b,c,, sono tutti poisitivi risulta $b^2=4ac$ sicchè sostituendo nella(1,2) troviamo gli autovalori :

$$x1,2=\frac{-b\pm\sqrt{b^2-b^2}}{2a}=\frac{-b}{2a}\quad \text{con } c=0\text{ , } x1=x2=\frac{-b}{2a} \quad (3)$$

Secondo caso:Se $\Delta= b^2-4ac>0$,Si trovano gli autovalori

$$x1=\frac{-b+\sqrt{b^2-4ac}}{2a}\quad(4)\text{ , }\qquad x2=\frac{-b-\sqrt{b^2-4ac}}{2a}\quad(5)$$

Introducendo questi valori nella(1) si verifica che per gli autovalori (3)due reali e conincidenti (4) e (5) reali e distinti Il **terzo caso** è risultato ostacolo insuperabile per gli algebristi del XVI secolo, impotenti quando il disciriminte $\Delta=b^2-4ac<0$,essendo i paramentri tutti positivi.Quindi il numero $\sqrt{b^2-4ac}$ non esiste.Nessun numero negativo ammette radice quadrata.Alle (4),(5)possiamo dare una inter- pretazione geometrica .Allo scopoconsiderando la funzione: $f(x)=ax^2+bx+c$ (7) La (7)nel piano x,y, se variano le coordinate del punto punto P(x,y) questo descrive uaa parabola ad asse verticave con la concavità rivolta verso il basso se i parametri a,b,c sono reali e positivi . Allo scopo consi- deriamo la funzione: $f(x)=ax^2+bx+c=y$ (8). Sappiamo che nei punti (4) e (5) si ha y=0.Come dire che la parabola(8) taglia l'asse x in **detti punti**. **Fig22-Amo** . Allora la (8),considerando variabili iparametri a,b.c e posto a=1 e posto la equazione: $x^2+px+q=0$ (9) con x variabile

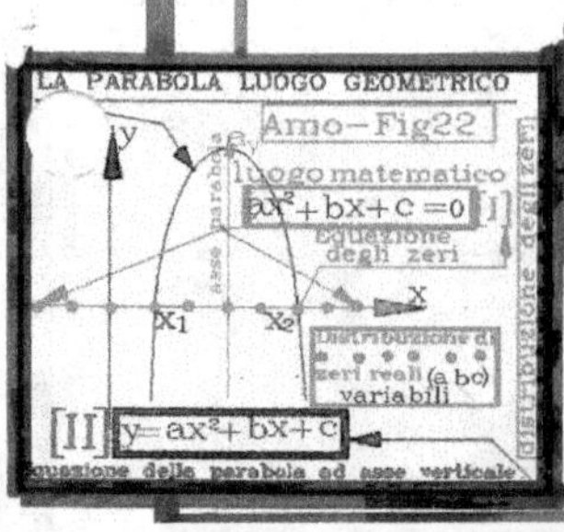

[I] The algebraic equations of the second degree :
Let $ax^2+bx+c=a(x-x_1)(x-x_2)=0$ (1) The first member , for **T.F.A**. ammette always two roots that satisfy it identically , that is, x_1 and x_2 . The problen is precisely the determination of the values of the roots or autovalori.Nella (1) identifies the number $\Delta = b^2- 4ac$ (2) said **discriminant** of the first member of (1) equivalent to the second member factorized . In general, the eigenvalues , according to theory are :

$$x_{1,2}=\frac{-b\pm\sqrt{b^2-4ac}}{2a}$$

(3) . The (3) have been solved my co- alg... equation of the third degree by Tartaglia, **Niccolò Fontana** (1499-1557) from wnich the deductible (2) . The solutions are diversified with the value of Δ.
First case : Let $\Delta = b^2- 4ac = 0$, then if the parameters a, b, c , , are all poisitivi is b2 = 4ac So that replacing in (1.2), we find the eigenvalues :

$$x1,2=\frac{-b\pm\sqrt{b^2-b^2}}{2a}=\frac{-b}{2a}\quad \text{con } c=0\text{ , } x1=x2=\frac{-b}{2a} \quad (3)$$

Second case: $\Delta= b^2-4ac > 0$, the eigenvalues are located :

$$x1=\frac{-b+\sqrt{b^2-4ac}}{2a}\quad(4)\text{ , }\qquad x2=\frac{-b-\sqrt{b^2-4ac}}{2a}\quad(5)$$

Introducing these values in (1) that occurs for the eigenvalues (3) two real and conincidenti (4)and (5 real and distinct The third case is the result insurmountable obstacle for algebraists of the sixteenth century , when the powerless disciriminte $\Delta = b^2- 4ac <0$, since all the parameters positive Therefor the number does not exist as the square roo of a negative number.To (4),(5)we can give ageome- tric interpretation tricity considering the function: $f(x) = ax^2 +bx +c$ (7) In the plane x,y, if you vary the coordinates of the point P (x,y) describes this uaa parable axis verticave with the concavity facing downwards if the parameters a, b, c are real and positive . In order consi - deriamo the function : $f(x)=ax^2 +bx+c =y$ (8) . We know that at the points (4) and (5) we have y = 0 . To say that the parable (8) cuts the x-axis in **these points. Fig22 -Amo** . Then (8),considering variables hhe parametri a,b,c , a=1and place to place and the equation $x^2+px+q= 0$ (9) with x variable

[I] SOLUZIONE GENERALE. Riprendiamo la nota equazione: $ax^2+bx+c=a(x-x_1)(x-x_2)=0$ (1) ,della quale conosciamo le soluzioni degli autovalori o radici del corpo dei numeri reali nei casi $\Delta=b^2-4ac\geq 0$ La gloria di aver creato il numero magico $j=\sqrt{-1}$ con il quale la fisica matematica sarebbe confinata nel deserto dei numeri reali spetta al grande matematico **Carl Friedrich Gauss**(1788-1855),ossia dopo circa due secoli dopo la soluzione della equazione(1) nel campo reale. Nella ipotesi : $\Delta=b^2-4ac<0$ (2)

[II] Proprietà algebriche.

$j=\sqrt{-1}$:

$j^2=\sqrt{-1}=-1$

$j^3=j^2j=-j$

$j^4=j^2j^2=-1\cdot-1=1$

Il problema di **Gauss** è stato quello di poter trasformare il discriminante $\Delta=b^2-4ac<0$, senza modificarne il valore. La geniale idea della unità immaginaria condusse al rapporto delle unità: $\frac{j}{j}=+1$ (3) che moltiplicata perla (2) non altera il - valore. Infatti : $(\frac{j}{j})^2(b^2-4ac)=(b^2-4ac)$ (4) .

[III] La soluzione del problema. Allo scopo le soluzioni degli autovalori nel caso di parametri a,b,c reali sono rispettivamente:

$$x_1=\frac{-b+\frac{j}{j}\sqrt{b^2-4ac}}{2a}=\frac{-b+j\sqrt{(b^2-4ac):j^2}}{2a}=\frac{-b+j\sqrt{(4ac-b^2)}}{2a}=z_1 \quad (5)$$

$$x_2=\frac{-b-\frac{j}{j}\sqrt{b^2-4ac}}{2a}=\frac{-b-j\sqrt{(b^2-4ac):j^2}}{2a}=\frac{-b-j\sqrt{(4ac-b^2)}}{2a}=z_2 \quad (6)$$

Le radici $z_1=\frac{-b+j\sqrt{(4ac-b^2)}}{2a}$ (7) e $z_2=\frac{-b-j\sqrt{(4ac-b^2)}}{2a}$ (8) complesse coniugate sono soluzione della equazione (1) dato che ora il discriminante $\Delta=4ac-b^2>0$. Non sappiamo esattamente come **Gauss** abbia scritto le varie operazioni sopra descritte ma il risultato finale è proprio quello od analogo alle soluzioni (7) ed (8).

Conclusione: $ax^2+bx+c=a(x-z_1)(x-z_2)$ (9)

[I] GENERAL SOLUTION . Resume the known equation : $ax^2+bx+c=a(x-x_1)(x-x_2)=0$ (1) , of which we know the solutions or roots of the eigenvalues of the body of the real numbers in the cases $\Delta=b^2-4ac\geq 0$ .The glory of having created the magic number $j=\sqrt{-1}$ with which mathematical physics would be confined to the desert of the real numbers it is up to the great mathematician **Carl Friedrich Gauss** (1788-1855) , ie after about two centuries after the solution of the equation (1) in the field real . In the case : $\Delta=b^2-4ac<0$ (2)

[II] Algebraic properties

$j=\sqrt{-1}$:

$j^2=\sqrt{-1}=-1$

$j^3=j^2j=-j$

$j^4=j^2j^2=-1\cdot-1=1$

The problem of Gauss was to be able to transform the discriminant $\Delta=b^2-4ac<0$, without changing the value. The brilliant idea of the imaginary unit led him to the ratio of the unity : $\frac{j}{j}=+1$ (3), which multiplied pearl (2) does not alter the - value .

In fact: $(\frac{j}{j})^2(b^2-4ac)=(b^2-4ac)$ (4) .

[III] The solution of the problem.

In order solutions of the eigenvalues in the case of parameters a, b , c are respectively real :

$$x_1=\frac{-b+\frac{j}{j}\sqrt{b^2-4ac}}{2a}=\frac{-b+j\sqrt{(b^2-4ac):j^2}}{2a}=\frac{-b+j\sqrt{(4ac-b^2)}}{2a}=z_1 \quad (5)$$

$$x_2=\frac{-b-\frac{j}{j}\sqrt{b^2-4ac}}{2a}=\frac{-b-j\sqrt{(b^2-4ac):j^2}}{2a}=\frac{-b-j\sqrt{(4ac-b^2)}}{2a}=z_2 \quad (6)$$

The roots $z_1=\frac{-b+j\sqrt{(4ac-b^2)}}{2a}$ (7) e $z_2=\frac{-b-j\sqrt{(4ac-b^2)}}{2a}$ (8)

are complex conjugate solution of the equation (1) since now the discriminant $\Delta=4ac-b^2>0$

We do not know exactly how **Gauss** wrote the various operations described above but the end result is just what or similar solutions (7) and (8).

Conclusion: $ax^2+bx+c=a(x-z_1)(x-z_2)$ (9)

La soluzione della((9)) in notazione complessa ha consentitio ai fisici di costruire un imponente edificio che ,senza soluzione i continuità, estende il dominio del sapere lo spazio geometrici allo spazio tempo. Basta cosdrare il quadrivettore spazio-tempo di **Albert Einstein** : $x_1^2+x_2^2+x_3^2=x_4^2=(jct)^2=-c^2t^2$ (1) ,essendo c la velocità della luce nel vuoto degli spazi interstellari postulata da Einstein. Come abbiamo già detto la luce è quel fenomeno di diffusione energetica luminiosa che si propaga per onde. E' stato Newton ad indagare sulla luce. Infatti, **Fig63-Fisot1 interpretò**

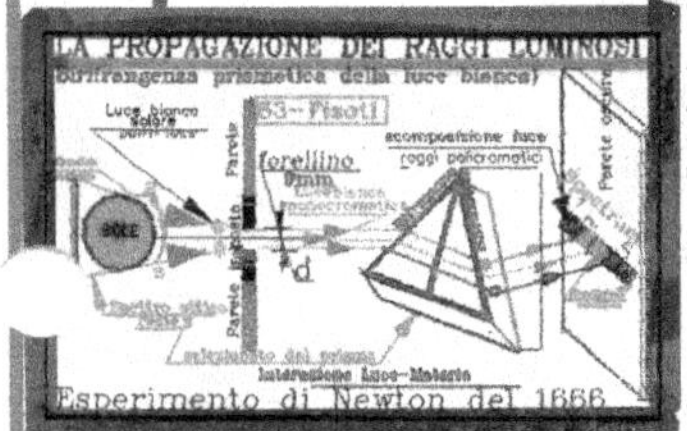

la diffrazione dell'onda luminosa come unaemmissione di particelle dermodinamiche che, grazie alla interposiziopne del prisma del cristallo poteva osservare lo spettro delle frequenze sotto forma dei colori. Dal rosso legato alla luce mediante la legge : $c=\lambda f$ (2) ; $f=\frac{c}{\lambda}$ (3) Come a suo dire c erano dei corpuscoli che come una qualiasi altra sorgente luminosa si diffondevano d'intorno . E' stato **Huygens** che, **Fig56-Fis4,** ha

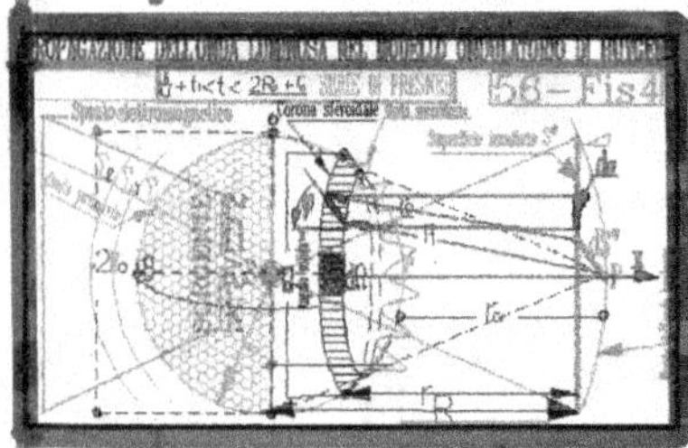

postulato che l'onda luminosa è una manifestazione di pura energia ottica. Infatti nella (Ap-Z/TavXIII) lo spettro ottico è di energia minima per il rosso di frequenza $f=\frac{c}{\lambda}=3{,}75.10^{14}$ Hz

max violetto $f=\frac{c}{\lambda}=7{,}50.10^{14}$ Hz **[I] LA FREQUENZA COME SEGNALE COSMICO** . La matematica pura è essenzialente un processo induttivo fine a se stesso ove non potesse trasformarsi in una sicura deduzione che fornisca una immagine del cosmo inteso come creazione dell'universo. Se ci soffermiamo sul concetto di frequenza abbiamo la consapevolezza che l'atimateria cioè la frequenza è il tratto di unione che completa l'universo. Con questo si intende dire che la frequenza, dimensionalmente inver- sa del tempo ,è la controparte fondamentale della fisica matematica postulata dalla (1) la cui radice è proprio la $j=\sqrt{-1}$ di **Gauss** . Vogliamo ora dare al capitolo una dimostrazione più concreta del problema diomstrando come si perviene ad una dimostrazione come una sintesi fra materia o energia frequenziale

The solution of ((9)) in complex notation has consentitio physicists to build an impressive building , the seamless continuity , extends the domain of knowledge space to space -time geome- try.Just cosdrare the four-vector space-time of **A. Einstein**: $x_1^2+x_2^2+x_3^2=x_4^2=(jct)^2=-c^2t^2$ (1) , c being the speed of light in vacuum of interstellar space postulated by Einstein. As we have already said, the light is the phenomenon of diffusion light energy that spreads waves. And ' Newton was to investigate the light.Indeed, Fig63-Fisot1 to inter- pret the diffraction of the light wave as unaemmissione particle dermodinamiche that , thanks to interposiziopne the prism of the crystal could observe the frequency spectrum in the form of colors. From the red bound by the law in the light : $c=\lambda f$ (2): $f=\frac{c}{\lambda}$ (3) . As he said that the corpuscles were c as a qualiasi other light source is spread round about . E ' was that **Huygens** , **Fig56 Fis4** , postulated that the light wave is a manifestation of pure optical energy . In fact, in (Ap-Z/TavXIII) the optical spectrum is minimum energy for the red frequency $f=\frac{c}{\lambda}=3{,}75.10^{14}$ Hz

and max violet $f=\frac{c}{\lambda}$ $7{,}50.10^{14}$ Hz

[I] HOW THE FREQUENCY SIGNAL COSMIC

Pure mathematics is essenzialente an inductive process if not an end in itself could become a safe deduction that provides an image of the cosmos as creation of the universe . If we focus on the concep of frequency we are aware that the atimateria ie the frequency is the stretch of union which completes the universe. By this we mean that the frequency , dimensionally -sa winter time , is the counterpart of the fundamental mathematical physics postulated by (1) the root of which is precisely the $j=\sqrt{-1}$ **Gauss** . To Want now give the chapter a more concrete demonstration of the problem diomstrando as you come to a demonstration as a synthesis of matter or energy frequential

La invenzione dei campi elettrromagnetici è opera del pensiero umano dedotto dalla reale esistenza della induzione elettromagnetica. Ci spieghiamo con un esempio. Quello che l'ago della calamita accerta puntando in una direzione è l'effetto induttivo che la terra intesa come un macromagnete con le sue polarità Nord(Zenit), Sud (Nadir) induce sulle opposte polarità dell'ago calamita. Fig10.Noto dall'antichità, in uso agli esattori del celeste impero che si orientavano su quel vasto territorio per riscuotere le imposte.

[I]Assiomatica dei campi{elettro-magnetici}↔{$\underline{k}$.$\underline{h}$}

La formulazione assiomatica dei campi in forma sistematica è di J.K.Maxwell(1831-1892) :

$\nabla\times\underline{k} = -\mu\frac{\partial \underline{h}}{\partial t}$ (1) , $\nabla\times\underline{h} = \underline{J} + \varepsilon\frac{\partial D}{\partial t} + \underline{J}i$ (2) .Leggenda:

$\nabla\equiv\frac{\partial}{\partial x}+\frac{\partial}{\partial y}+\frac{\partial}{\partial z}$ (3) Operatore alle derivate spaziali di punto variabile nello spazio di permittività ma magnetica μ ed elettrica ε ,essendo $\underline{J}$ la corrente elettrica libera nello spazio e $\underline{J}i$ la corrente elettrica imposta a mezzo opportuno generatore .La (1) si legge< Il prodotto del vettore campo elettrico è proporzionale alla permitività magnetica cambiata di segno per la derivata rispetto al tempo del campo magnetico> Per la (2) si leggerà < Il prodotto del vettore del campo magnetico è proporzionale alla permittività elettrica per il vettore spostamento elettrico derivata rispetto al tempo associata alle eventuali correnti elettriche libere nello spazio od imposte > E' fondamentale riconoscere che la(1) come la (2) hanno in comune la dipendenza dal tempo istantaneo t e sfrondata dalle correnti la (2),si riducono al sistema: $\nabla\times\underline{k} = -\mu\frac{\partial \underline{h}}{\partial t}$ (I) , $\nabla\times\underline{h} = \varepsilon\frac{\partial D}{\partial t}$ (II). Sistema di due{$\underline{k}$.$\underline{h}$}incognite e un unico termine noto $\underline{D}$ .I primi membri sono detti rotori {$\underline{k}$.$\underline{h}$} . Per le correnti si parla di divergenze: $\nabla\cdot(\underline{J}i+\underline{J}) = -\frac{\partial\rho}{\partial t}$ (4) con ρ densità.Le equazioni precedenti in ℜ sono valide anche in campo ℭ ,cioè nel dominio dei vettori complessi traducibili in vettori reali e o complessi:

- $\tilde{e}(t) = Em\sin(\omega t+\phi) \in \Re\ \dot{E} = |Em|\, e^{j(\omega t+\phi)}$ (5)

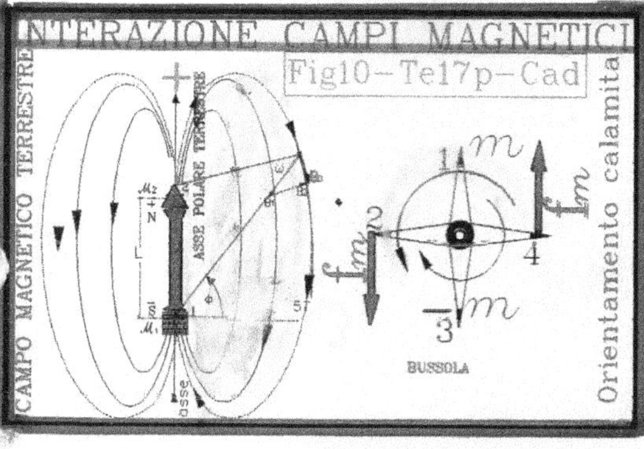

The invention of the fields elettrromagnetici is the work of human thought deducted from the actual existence of electromagnetic induction. There to spread with an example. That ensures that the needle of the magnet pointing in one direction is the inductive effect that the land intended as a macromagnete with its polar -ity North (Zenit), South (Nadir) induces on the opposite po- larity of the needle magnet. Fig10.Noto ancient times, used to the debt of the celestial empire that would orient themselves on the vast territory to collect taxes .

[I] Axiomatic of {electro- magnetic **fields**}↔{$\underline{k}$.$\underline{h}$}

The axiomatic formulation of the fields in the **form** of systematic **J.K.Maxwell** (1831-1892) :

$\nabla\times\underline{k} = -\mu\frac{\partial \underline{h}}{\partial t}$ (1) , $\nabla\times\underline{h} = \underline{J} + \varepsilon\frac{\partial D}{\partial t} + \underline{J}i$ (2). Legend:

$\nabla\equiv\frac{\partial}{\partial x}+\frac{\partial}{\partial y}+\frac{\partial}{\partial z}$ (3) Operator to the spatial derivatives of variable point in the space of permissible magnetic μ and electrical and ε , being the electrical current $\underline{J}$ free space and the electric current $\underline{J}i$ of generator should be set to half . The(1) reads < The product of the electric field vector is proportional to the magnetic permitività changed in sign for the time derivative of the magnetic field > for the (2) is read < the product of the vector of the magnetic field is proportional to the dielectric constant for the displacement vector electric derivative with respect to time associated with any electrical currents freely in space or taxes > It ' important to recognize that (1) like (2) have in common the time dependence instant you that , stripped from the current (2) , are reduced to the system: $\nabla\times\underline{k} = -\mu\frac{\partial \underline{h}}{\partial t}$ (I), $\nabla\times\underline{h} = \varepsilon\frac{\partial D}{\partial t}$ (II) System of two {$\underline{k}$.$\underline{h}$ } unknowns and a single term known $\underline{D}$ .The first members are rotors {$\underline{k}$.$\underline{h}$ } . For currents are talking about diverences : $\nabla\cdot(\underline{J}i+\underline{J}) = -\frac{\partial\rho}{\partial t}$ (4) wich the density . ℜ in the above equations are also valid in the ℭ field , that is in the domain of the complex vectors translatable in real vectors co complexes :

- $\tilde{e}(t) = Em\sin(\omega t+\phi) \in \Re\ \dot{E} = |Em|\, e^{j(\omega t+\phi)}$ (5)

J.K. Maxwell (1831-1892)

Per una utile comparazione ,partendo dalla equazione di continuità: $\nabla\cdot(\underline{J}i+\underline{J})=-\frac{\partial\rho}{\partial t}$ (4) di natura assiomatica relativa ai campi {k̲ .h̲} , supportiamo in sede sperimentale la creazione reale degli stessi. Nella **Fig18** è rappresentato un oscillatore ad una maglia con in serie una impedenza : $\dot{Z}=R+j(\omega l-\frac{1}{\omega C})$

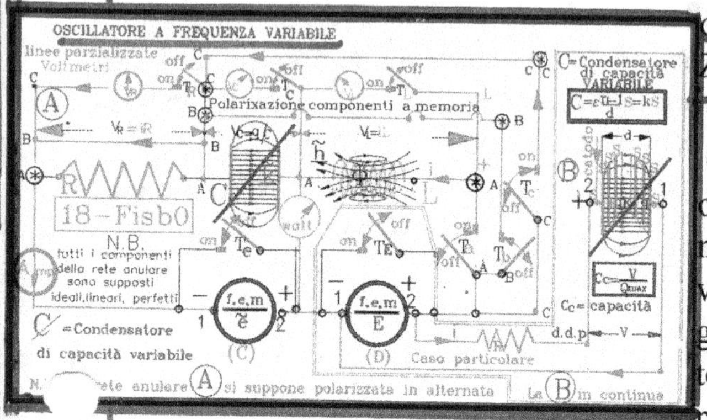

Come si può osservare la maglia è attivata da due generatori di tensione ed è polarizzabile usando gli operatori **T**(interruttori) **on line**. Ora si supponga all'istante to=0 di aver chiuso il generatore **(C)** e Ta .La combinazione del campo magnetico h̲ della induttanza L per effetto della corrente i(t) è tale da associarsi al campo elettrico k̲ , per ragioni di continuità giusto la (4).Nella quale è veicolata la densità ρ di carica di elettroni, espressa dalla equazione in corrente istantanea: $i(t)=\nabla\cdot(\underline{J}i+\underline{J})=-\frac{\partial\rho}{\partial t}\equiv\underline{J}i$ (5)

(J̲i corrente impressa dal generatore **(C)**). La J̲ libera fluisce dal condensatore**C** all'utilizzatore **R** od altro.Componenti ideali e lineari costanti nel tempo. Ma l'oscillatore è anche una sorgente di onde {k̲ , h̲ } che si diffondono nel mezzo{ε,μ}o nel vuoto{εo,μo} non si limita ad operare le descritte funzioni ma emette delle onde di energia di cui stiamo dando una immagine assiomatica dovuta ad **H.Helmholtz**(1821-1894).Le onde in unmezzo lineare ed isotropo, invariante nel tempo,si formalizzano partendo dalle equazioni di **Maxwell**: $\nabla\times\underline{k}=-\mu\frac{\partial\underline{h}}{\partial t}$ (I), $\nabla\times\underline{h}=\varepsilon\frac{\partial\underline{D}}{\partial t}+\gamma\underline{D}+\underline{J}i$ (II) **[I] Si tratta di un sistema non a variaili separabili di equazioni differenziali alle derivate parziali del primo ordine.**

[II] Soluzione del sistema Il sistema (I),(II) **consiste nel separare le incognie in modo tale da ottenre una equazione differenziale con una unica grandezza di campo** La teoria della integrazione insegna che è possibile ma che nello sstesso tempo la equazione differenzaile trasformata in tal modo presenta una sola incognità ma elevata di un ordine rispetto alla iniziale.

For a useful comparison , starting from the continuity equation : $\nabla\cdot(\underline{J}i+\underline{J})=-\frac{\partial\rho}{\partial t}$ (4) of axiomatic nature relating to the fields {k̲ .h̲ } , We support in the experimental the actual creation of the same. In **Fig18** shows an oscillator with a mesh in a se-ries impedance : $\dot{Z}=R+j(\omega l-\frac{1}{\omega C})$

As you can see the mesh is activated by two voltage sources and is polarizable using the operators T (switches) **on line** . Now suppose the instant to = 0 have closed the generator **(C)** and Ta .

The combination of the magnetic field of the indut-tance L due to the current i(t) is such as to associate with the magnetic h̲ field , for reasons of continuity the right (4) . Which is conveyed in the density ρ of charge of the electrons , expressed by the equation in courrent instantaneous : $i(t)=\nabla\cdot(\underline{J}i+\underline{J})=-\frac{\partial\rho}{\partial t}\equiv\underline{J}i$ (5) (the J̲i impressed current from the generator **(C)**) . The J̲ free ted flows from the user condensatore **C** a to utilizer **R** or other component ideal linear and constant over time.

But the oscillator is also a source of waves {k̲ , h̲ } that spread in the middle{ε, μ}or in vacuum{εo, μo} , is not limited to operate descrcipt to describe func-tions but emits electromagnetic waves of energy that we are giving an image axiomatic due to **H Helholtz** (1821-1894) . Waves in a middle linear and isotropic, time-invariant , are formalized starting from **Maxwell's** equations :

[I] $\nabla\times\underline{k}=-\mu\frac{\partial\underline{h}}{\partial t}$ (I) , $\nabla\times\underline{h}=\varepsilon\frac{\partial\underline{D}}{\partial t}+\gamma\underline{D}+\underline{J}i$ (II)

This is a non- separable variaili of partial differential equations of the first order.

[II] Solution of the system The system (I) , (II) **consists in separating the unknown in manner such a way that a differential equation with a unique greatness of field**

The theory teaches that integration is possible but that at the time same equation differenzaile transformed in this way presents only one unknown but of a high order with respect to initial .

Con riferimento all'oscillatore di **Fig16-Te16** ed alla $\dot{Z}$ impedenza: $\dot{Z} = R + j(L\omega - \frac{1}{\omega C})$ (1) - Nicchia **(a)**

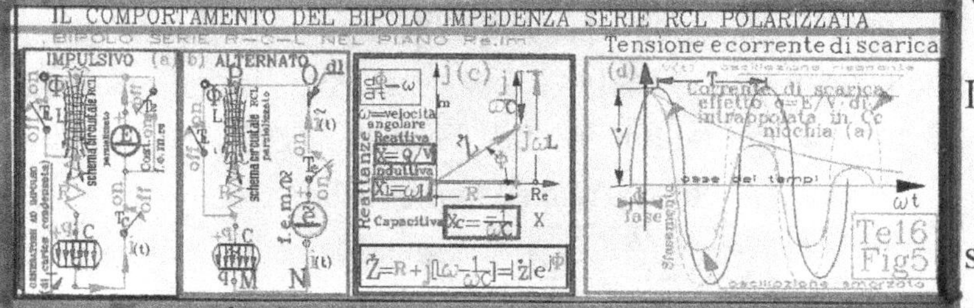

campo$\{\underline{k}, \underline{h}\}$ elettromagnetico **(b)**. E' legato al condensatore **C dalla pulsazione** ω. Se per $t=t_o$ si è caricato a C_C polarizzando, la (1) della serie RCL in **linea** .In condizione di equilibrio in tensione corrente si ha $V: V_C+V_L+V_R=e(t)$ (2) indotta dalla f.e.m : $e(t)=E_m \sin(\omega t+\phi)$ (3),rappresentazione grafica della impedenza nella nicchia **(c)**

[I] La soluzione in corrente alternata(**nicchia(b)**)
Se si chiude nell'istante $t=t_o=0$ **Te** (lato O-N) si polarizza la impedenza(lato P-M) che per effetto (3) impone una corrente $i(t)$. Nel caso di una sola maglia quale l'oscillatore è facile da risolvere in tempo reale.Allo scopo(2)trasformata nelle componenti: $\{V_C=q/C (2_1), d\Phi/dt=i'L (2_2), [F'=Li'+iL'](2_{2'})\}$.Ma supposto: L=Cost è iL'=0], allora : $V_R=iR$ (2_3) (4)
Le(4)cosentono di trasformare la equazione (2) nella dinamica: $q/C+i'L+iR=e(t)$ (2') che derivata : $\frac{d}{dt}[q/C+i'L+iR=e(t)]=\frac{i}{C}+i''L+i'R=\omega E_m\cos(\omega t+\phi)$ (5)
Teorema per il caso (5) < La soluzione generale y(t) di una equazione differenziale del 2° ordine è uguale alla somma della sua omogenea $y_o(t)$più una associata particolare $y_p(t)$>**Formalizzato** $y(t)=y_o(t)+y_p(t)$(6)
[I]Soluzione di $y_o(t)$.Allora da: $i/C+i''L+i'R=0$ (7) inmodalità : $i''+(R/L)i'+(1/CL)i = i''+\sigma i'+\tau=0$ (8) ,con $\sigma=\frac{R}{L}$ costante dissipativa e τ costante di tempo, connessa alla pulsazione o frequenza: $f=\omega_o/2\pi$ (f)
La soluzione della (8) implica la detrminazioni dei suoi autovalori definiti dalla sua equazione caratteristica ottenibile dalla (8) sostituendo le derivate con il fattore λ .Quindi : (9) . Si tratta di una equazione tipo $\lambda^2+\sigma\lambda+\tau=0$ di 2° grado con radici:
$\lambda_1=[-(R/L)+j\sqrt{(R/L)^2-4/CL}\ ]/2$ (10)
$\lambda_2=[-(R/L)-j\sqrt{(R/L)^2-4/CL}\ ]/2$ (11). Questi autovalori consentono di definire le soluzioni esponenziali : $e^{\lambda 1t}$(I), $e^{-\lambda 2t}$(II) particolari della (8) (a radici complesse coniugate Infatti per il **T.F.A**. in quanto tutte le componenti sono positive)

With reference oscillator **of Fig16 - TE16** and the impedance: $\dot{Z} = R + j(L\omega - \frac{1}{\omega C})$ (1) - Niche **(a)** field $\{\underline{k}, \underline{h}\}$ electromagnetic **(b)**

Is therefore linked to the capacitor **C by pulsation** ω

If for $t = t_o$ is loaded C_C biasing , the (1) of the series RCL **on line** .

In equilibrium condition in voltage current one has $V: V_C+V_L+V_R=e(t)$ (2) induced e.m.f : $e(t)=E_m \sin(\omega t+\phi)$ (3) , graphical representation of the impedance in the niche **(c)**

[I] the solution in alternating current (**niche (b)**)
If you close the instant t = to **Te** = 0 (O-N side) polarizes the impedance (P-M side) and due to (3) requires a current $i(t)$. In the case of a single mesh as the oscillator in question is quite easy to solve the problem in real time. At the end, the (2) is transformed with the equations of the components: $\{V_C=q/C (2_1), d\Phi/dt=i'L(2_2)[\Phi'=Li'+iL'](2_{2'})\}$. But supposed [L Cost = iL'= 0], is $V_R= iR$ (2_3) (4)
The(4) to alows transform equation (2) in the dynamics : $q/C+i'L+iR=e(t)$ (2').To derived that : $\frac{d}{dt}[q/C+i'L+iR=e(t)]=\frac{i}{C}+i''L+i'R=\omega E_m\cos(\omega t+\phi)$ (5)

Theorem for the case (5) <the general solution y (t) of a differential equation of second order is equal to the sum of its homogeneous $y_o(t)$plus an associated ta particular $y_p(t)$ >**Formalized** $y(t)=y_o(t)+y_p(t)$(6)
[I] solution $y_o(t)$. then by $i/C+i''L+i'R=0$ (7) mode the $i''+(R/L)i'+(1/CL)i = i''+\sigma i'+\tau=0$(8) , with σ= R/L constant and τ dissipative time constant , which is connected to the pulse or frequency : $f=\omega_o/2\pi$. (f) **The solution** of (8) implies the determination of its eigenvalues defined by its characteristic equation obtainable from (8) by replacing the derivatives with the factor λ. So: $\lambda^2+\sigma\lambda+\tau=0$ of 2° degrè wich eingevalues:
$\lambda_1=[-(R/L)+j\sqrt{(R/L)^2-4/CL}\ ]/2$ (10)
$\lambda_2= [-(R/L)-j\sqrt{(R/L)^2-4/CL}\ ]/2$ (11)
These allow you to define the exponential solutions : $e^{\lambda 1t}$(I) , $e^{-\lambda 2t}$(II) details of (8) (a complex coniugate roots for.The fact T.F.A.in that all the components are positive)

Come a pag-53,la unità immagnaria j ha introdotto un supporto fondamenatele nella geometria euclidea, estendendo le applicazioni nelle quali interviene non solo spazio ma anche il tempo. Lo stesso Einstein postula la esistenza di una comtropparte fisica legata alla velocità scalare della luce: $c = \lambda f = Cost$ (I) di un quadrivettore(4) mai [prima di allora (1905 ?)pensato]. Ideato in base al principio che la c è,nel vuoto,costante rispetto al tempo.

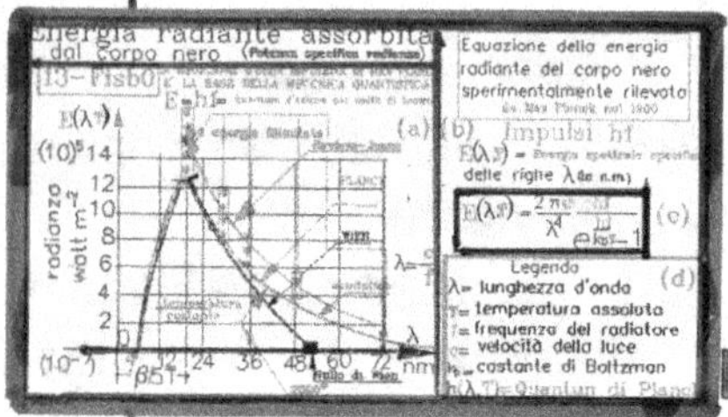

[I] Occorre precisare che l'onda luminosa, Fig13, si presenta anche con carattere impulsivo**Max Planck**(1858-1947)nella meccanica ondulatoria: $dE=hf$(1) con **h**(quntum **impulsivo**)

[II]Come interviene l'algebra complessa di **Gauss** nella (1) e (4)?La legge che lega λ(lunghezza d'onda) alla frequenza $f=\frac{c}{\lambda}$(2) esprime la velocità del vettore spazio nel tempo t: $d=ct$(3).Quindi la frequenza f se c è costante saràelevata quanto minore è λ. Infatti la equazione $f\lambda=c$=Cost conferma il comportamento(4)

[III]A.Einstein nella ipotesi sopraesposte ha postulato: $x_1^2+x_2^2+x_3^2=-x_4^2$(2) , cioè il vettore scalare geometro in c.c.c. del primo membro al vettore fisico $-x_4^2$. Il problema consiste nel dimostrare come la distanza geometria $d^2=x_1^2+x_2^2+x_3^2$ possa essere tradotta in senso fisico. Allo scopo $d^2=x_4^2=c^2t^2$ (21), che introdotta nella (2) dà: $x_1^2+x_2^2+x_3^2=-c^2t^2$ (22) ,oppure : $d=\sqrt{-c^2t^2}$? Non esiste la radice quadrata di un numero negativo . La soluzione usata da Einstein è stata suggerita da Gaus, pg53, per cui il quadrivettore (2) si con figura $x_1^2+x_2^2+x_3^2=-(j^2c^2t^2)=+c^2t^2$ (3)

[IV] L'effetto fotoelettrico La Fig10-Fis1 mostra come un flusso di fotoni **hf** incidendo su un cristallo al Selenio per urto con le particelle eletroni liberi del cristallo possso fungere da f.e.m e generare la corrente i di conduzione come mostra il collegamentone della nicchia (a).Il tuttto nella pila (b) e l'effetto fotoelettrico i in lux,nicchia (c)

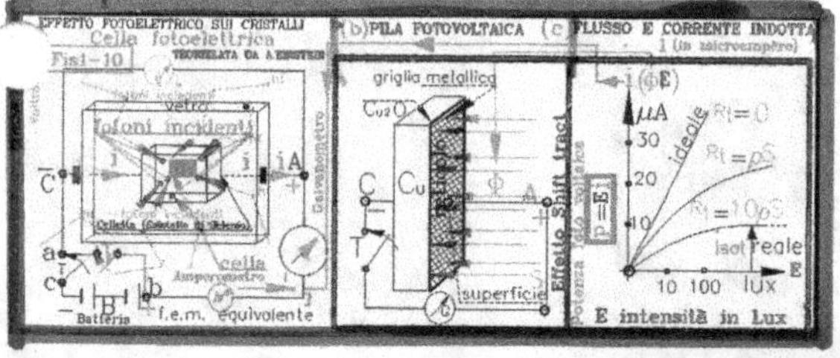

How on page 53, the unit $j=\sqrt{-1}$ immagnaria has introduced a support fondamenatele in Euclidean geometry , extending applications in which intervenes not know - space but also time . Einstein himself the postu - the existence of a comtropparte physics related to the scalar velocity of light: $c = \lambda f = Cost$ (I) a four (4) never [before then (1905 ?) Thought] . Designed according to the principle that the c is , in a vacuum, constant with respect to time .

[I] should be noted that the wawe leight , Fig13 , is also character impulsivo**Max Planck** (858-1947) in wave mechanics: $dE=hf$ (1) with h (quntum impulsive)

[II] As intervenes algebra complex **Gauss** in (1) and (4)? The law which binds (wavelength) at frequency $f=\frac{c}{\lambda}$ (2)expresses the speed of the vector space in time t : $d = ct$ (3). So if the frequency f e c is constant , the lower the saràelevata . In fact, the equation $f\lambda=c$=Cost . confirmation behavior (4)

[III]A.Einstein postulated in the hypothesis above show: $x_1^2+x_2^2+x_3^2=-x_4^2$ (2) ie the vector scalar scalar wiht carrier physical - x_4^2 . The problem is to show how the geometry distance: $d^2=x_1^2+x_2^2+x_3^2$ can be , translated in a physical sense. In order $d^2=x_4^2=c^2t^2$ (21) , who introduced in (2) gives : $x_1^2+x_2^2+x_3^2=-c^2t^2$ (22), or $d=\sqrt{-c^2t^2}$ = ? There is the square root of a negative number.The solution used by **Einstein** was suggested by **Gaus** , pg53 , for which the four-vector (2) is to figure $x_1^2+x_2^2+x_3^2=-(j^2c^2t^2)=+c^2t^2$ (3)

[IV] The photoelectric effect The Fig10 - Fis1 shows how a stream of photons **hf** etching on a crystal to Selenium for collision with the free electrons of the crystal particles to serve from **emf** and generate the courrent i of the condution niche (a) . La battery alla wich in on to bear courrent of conduction . Pile tuttto the guarded in (b) and the photoelectric effect in lux (c) in power

M. Planck (1858-1947)

A. Einstein (1879-1955)

Questo capitolo è alla base della risoluzione delle reti elettriche analogiche a costanti R-C-L .

P1- Operatore di Laplace.

Si indica con la notazione simbolica $\nabla^2 \psi = 0$ (1) la equazione differenziale del secondo ordine alle derivate parziali nelle variabili x,y della funzione continua e a derivate seconde non nulle ,definita:

$$\frac{\partial^2\Psi}{\partial^2 x} + \frac{\partial^2\Psi}{\partial^2 y} = 0 \equiv \nabla^2\Psi = 0 \quad [4].$$

nella quale la equazione $\nabla^2 = \frac{\partial^2}{\partial^2 x} + \frac{\partial^2}{\partial^2 y}$ [5]

prende il nume di operatore differenziale laplaciano

P2-Definizione della trasforrmata di Laplace

Se F(t) è una funzione definiita per valori di t>0 , e indichiamo la trasformata di laplace con LT{f(t)} e con F(s) la sua trasformata si scrive, per definizione:

$$LT[f(t)] = F(s) = \int_0^{+\infty} e^{-st} f(t)\, dt$$

(6) nella quale si indica l'esponente $s = \sigma + j\omega t$ (a) complesso

P3- Nella risoluzione delle reti elettriche (R.E.)

Supposto che il cirduito o maglia della rete , con un ingresso u(t)cisoidale, tipo: $e(t)=E_m \sin(\omega t+\alpha)$ (7) a costanti concentrate R-C-L indipendenti dal tempo Allora assumono un significato elettrofisico Dato che $\sigma = \frac{R}{L}$ = (b)= dissipazione energia fornita dalla (7) , $\omega = 2\pi f = \frac{1}{LC}$ (c)=pulsazione ciclica o frequenza, R espresso in unità ohm, f espresso in Hz

P4- La trasformazione inversa della (6) Detta anche **antitraformata di Laplace** assume la forma

$$LT^{-1}[F(s)] = f(t) = \frac{1}{2j\pi} \cdot \int_{a\to -j\infty}^{a\to j+\infty} e^{st} F(s)\, ds \quad (8)$$

P5- Le trasformate delle serie di Fourier

Consideriamo la serie monolatera di Fourier e la famiglia di funzione $\{U_k\}_{k\in n=1,2,..}$ .tali che converga la serie $F(s) = \sum_{n=0}^{+\infty} C_n U_n$ (9) .Si può allora dire : << che la F(s) è sviluppabile in serie di $\{U_k\}_{k\in N}$>> Con analogo criterio mutando nella (9) $+\infty$ in $-\infty$ si

This chapter is the basis of the resolution of the analog electrical networks in constant R-C-L

P1-Laplace operator.

It indicates the symbolic notation: $\nabla^2 \psi = 0$ (1) equation the second order differential in the variables x, y of the continuous function and non 0, **in**

$$\frac{\partial^2\Psi}{\partial^2 x} + \frac{\partial^2\Psi}{\partial^2 y} = 0 \equiv \nabla^2 Y = 0$$

[4], derivatives defined in which the equation : $\nabla^2 = \frac{\partial^2}{\partial^2 x} + \frac{\partial^2}{\partial^2 y}$ [5]

takes the nume of differential Laplacian operator

P2-Definition of trasforrmata Laplace

If F (t) is a function definiita for values of t > 0, and denote the Laplace transform with LT {f (t)} and F (s) transformed his writing, by definition:

$$LT[f(t)] = F(s) = \int_0^{+\infty} e^{-st} f(t)\, dt$$

(6) which ndicate the complex exponent $s = \sigma + j\omega t$ (a)

P3-Resolution grids eletrical (RE)

Assuming that the circuit or mesh network , with an input u(t)cisoidal, type: $e(t)=E_m \sin(\omega t+\alpha)$ (7) and constant R-C-L independent of time, then take on a meaning electrophysical since $\sigma = \frac{R}{L}$ (b) = dissipation energy given by (7), $\omega = 2\pi f = \frac{1}{LC}$ (c) = cyclic or pulsation frequency, expressed in units of R ohms, f expressed in Hz

P4-the inverse transformation of (6)

Also called antitraformate of Laplace assumes the form inverse transformation of (6) form

$$LT^{-1}[F(s)] = f(t) = \frac{1}{2j\pi} \cdot \int_{a\to -j\infty}^{a\to j+\infty} e^{st} F(s)\, ds \quad (8)$$

P5- the transformed of Fourier series

Consider the **Fourier** series monolatera and family of function $\{U_k\}_{k\in n=1,2,...}$ converge such that the series $F(s) = \sum_{n=0}^{+\infty} C_n U_n$ (9.) We can then say:<< That F (s) is developable in series of $\{U_k\}$ kN>> With analogous criterion .to obtien the series inverse mutating (9) $+\infty$ in $-\infty$

P_1- Funzioni di ordine esponenziale

Se esiste, per due costant M>0 e γ i la condizione $|e^{-\gamma t}f(t)|<M$ (1) si dice f(t) esponenziale di ordine γ Es Se γ-4 e si pone $f(t)=e^{5t}$ risulta per la (1) . $e^{5t} > e^{4t}$ dato che 5t>4t L'esponenziale di ordine 5 per $f(t)=e^{4t}$ soddisfa la condizone (1)

P_2- Alcuni esempi di trasformate di Laplace

Ricordiamo che per definizione la trasformata di Laplace è : $LT\{f(t)\}=f(s)=\int_0^{+\infty} e^{-st} f(t)dt$ (2)

in cui supponiamo per ora che il parametro s sia reale La f(t) è laplace trasformabile se la (2) converge su tutto l'asse reale , Fig 1(a). Riportiamo di seguito le trasformate di qualche funzione reale tale che $f(t) \leftrightarrow f(s)$ (3) Tab-a.Am13 Come si potrà nel seguito costatare, le funzioni trigonometriche sono fondamentali nel capitolo(R.E) .

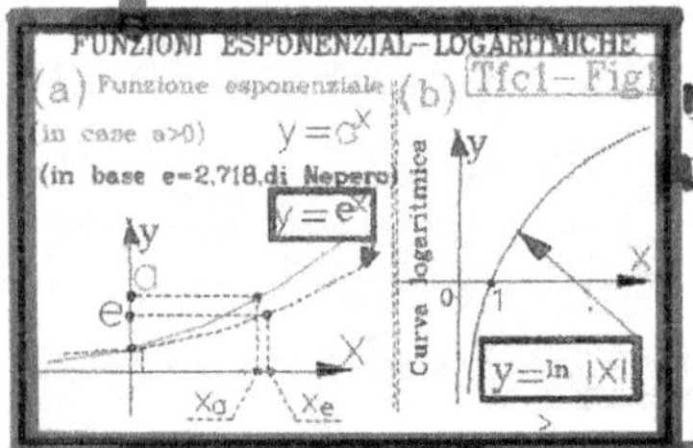

Possiamo dedurre la trasformata:

$$LT[tn(at)]=LT[\frac{a}{s^2+a^2}\frac{s^2+a^2}{s}]=\frac{a}{s} \quad (4)$$

usando l'operatore (2): In generale le trasformazioni con la (2) sono laboriosa per cui il dipartimento di fisica matematica applicata ha costruito dei tabulat LT distinte con le lettere A-B-C-D-E-F Introdotte dalla presentazione L di Laplace Per comple-completare la statistica le derivate delle f(t)=f(s) e gli integrali delle F(t)=f(t) Per il completamento della casistica ci colleghiamo alle funzioni di variabile complessa z con le derivate fatte rispetto alla z variabile . Si tratta di funzioni cosidette analitiche derivabili Abbiamo preparato il terreno per iniziare la risoluzione delle R.E. usando il metodo simbolico che è strada meno impervia delle soluzioni di R.E.per mezzo di equazioni differenziali

TRAFORMATE DI LAPALCE DI FUNZIONI ELEMENTARI
{f(t)} Tab-a/Am13 LT{f(t)} = f(s)

1	1		$\frac{1}{s}$	s>0
2	t		$\frac{1}{s^2}$	s>0
3	t^n		$\frac{n!}{s^{n+1}}$	s>0
4	e^{at}		$\frac{1}{s-a}$	s>a
5	sin(ta)		$\frac{a}{s^2+a^2}$	s>a
6	cos(ta)		$\frac{s}{s^2+a^2}$	s>0
	sinh(ta)		$\frac{a}{s^2-a^2}$	s>|a|
8	cosh(ta)		$\frac{s}{s^2-a^2}$	s>|a|

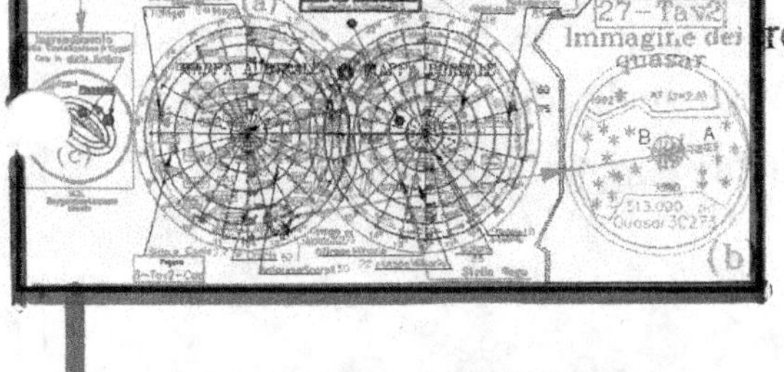

P_1-order functions exponenzial

If it exists, for two costant M> 0 and γ the condition $|e^{-\gamma t}f(t)|<M$ (1) is said to f (t) of exponential order γ Es If γ-4 and arises $f(t) = e^{5t}$ is for (1).. $e^{5t} > e^{4t}$ since 5t> 4t the exponential order 5 for $f(t) = e^{4t}$ satisfies the condizone (1)

P_2-Some examples of Laplace transforms

Recall that by definition the Laplace transform is :

$$LT\{f(t)\}=f(s)=\int_0^{+\infty} e^{-st} f(t)dt \quad (2)$$

in which we assume for now that the parameter s is real the f (t) is Laplace transformable if the (2) conver-ge of all the real axis , Fig 1 (a). Here be-low the transform of some real function such that $f(t) \leftrightarrow f(s)$ (3) Tab-a.Am13

As can be followed in the establishment, the trigonometric functions are critical in the chapter (RE).

We can deduce the transform:

$$LT[tn(at)]=LT[\frac{a}{s^2+a^2}\frac{s^2+a^2}{s}]=\frac{a}{s} \quad (4)$$

using the operator (2):

In general, the transformations with (2) are laborious to which the physics department of applied mathematics has constructed of Tabulat LT distinct with the letters A-B-C-D-E-F Introdottedalla presentation of L **Laplace** to complete the statistical comple-the derivatives of f(t) = f (s) and the integral of F(t) = f (t) for the completion of cases we connect to the functions of a complex variable z-derivative made with respect to the variable z. This is the so-called analytic functions derivable Rato We prepare the ground to start the resolution of R.E. using the symbolic method which is way less impervious solu tions of Reper means of differential equations

P_1- Resistore-Condensatore-Induttore

Questi componenti posti in serie fra loro costituiscono (On Line) in linea una impedenza $\dot{Z}$ è completa dei componenti analogici : $\dot{Z}=R+j[L\omega - \frac{1}{\omega_, C}]$ (1) Nella

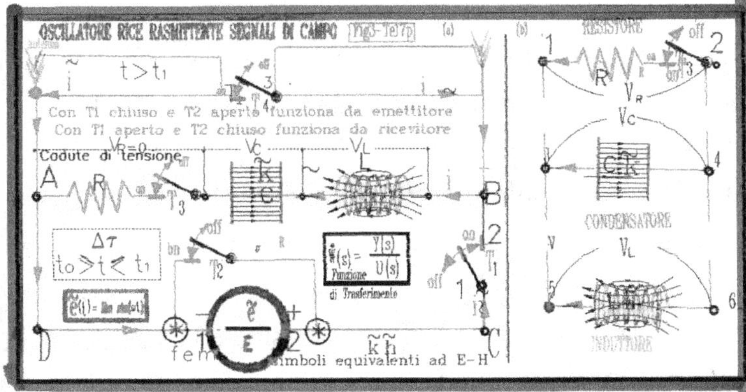

Fig3 nicchia (a) la $\dot{Z}$ è collegata in serie ai morsetti A e B con la forza elettromotrice, fem formalizzata f.e.m. : $\tilde{e}(t)=Em \sin(\omega t + \phi)$ (2) .Nella quale w e la pulsazione applicate al circuito al momento della chiusura dell'interruttore T_1

P_2- Il comportamento circuitale .

Se nell'stante t=0 si chiude T_1 (posizione on line) possono verificarsi delle uscite y(t) per t≥0 dipendenti da diverse cause. Precisamente :

1- I componenti a memoria possono essere dello stato zero (-t<0) oppure potenzialmete attivo .

2 - Se l'interruttore T_2 è aperto (off line) la corrente non può circolare e il segnale dell'ingresso u(t) costitituito dalla (2) non può inviare segnali attraverso Z.

3- Nella ipotesi 2 se il condensatore non è nello stato zero, cioè ha intrappolato una vcarica elettrostatica **q** allora si ha una uscita y(t) in corrente transitoria.

4. Si supponga che i compoenti a memoria siano nello stato zero e che per t=0 si chiuda l'interruttore T_1, tutti gli altri aperti.

E' questo il caso del funzionamento continuo del circuito oscillante a spese della **f.e.m**. carattere permanente se non si apre T_1,

P_3-Il comportamento circuitale di singoli elementi.

Nella nicchia di Fig3(b) abbiamo separato i componenti R-C-L per una rapida analisi del loro funzionamento sia in tempo reale t che virtuale s .

P_1- Resistor- Capacitor- Inductor

These components, placed i n series together constitute(On Line)in a line impedance is the complete analog components : $\dot{Z}= R + j [L\omega- \frac{1}{\omega_, C}]$ (1)

In the niche Fig3 (a) $\dot{Z}$ is connected in series to the terminals A and B with the electromotive forc fem formalized : f.e.m. : $\tilde{e}(t)=Em \sin(\omega t + \phi)$ (2) Which we In the pulsation applied to the circuit at the time of closure of the switch T_1

P2- The circuit behavior.

If nell'stante t = 0 closes T1 (position online) can occur output y (t) for t0 employees from different causes. precisely:

1 - The components in the memory can be state zero (-t <0) or potenzialmete active.

2 - If the switch T_2 is open (off-line), the current can not circulate and the input signal u (t) to make by (2) can not send signals through Z.

3 - In case 2, if the capacitor is not in was zero, ie has trapped a vcarica electrostatic **q** then we have an output y (t) in the transient current

4. Assume that any components in memory are in the zero state and that for t = 0 closes the switch , T_1 all the others open. And 'This is the case of the continuous operation of the oscillator at the expense of the .. permanent if I will not open T_1

P3- The circuit behavior of individual items.

In the niche of Fig3 (b) we have separated the components R-C-L for rapid analysis of their operation both in real time t and virtual s.

E utile analizzare il comportamnto di ogni singolo elemento dei circuiti analogici ,sia nelle condizioni statiche, invarianti rispetto al tempo , o variabili in temo reale . Spratutto ,ai fini applicativi , variabili in tempo virtuale (s) La fig2- Te17 mostra la seguenza dei componenti separabili mediane interruttori.

I- IL RESISTORE .Questo componente è definito come elemento passivo delle **R.E**. E' privo di memoria elettromagnetica. Nel senso che tutta la energia erogata dalla f.e.m. di polarizzazione della impedenza $\boxed{Z = R + j[L\omega - \frac{1}{\omega C}]}$ (1) è dissipata, effetto Joule, da R. Alle alte frequenze il comportamento di R non è tale da associare alla dissipazione un campo elettromagnetico {$\underline{k}$, $\underline{\hbar}$} . Questo perchè la materia cristallina dei resistori tende a convogliare gli elettroni del mantello nucleare sicchè crescendo in frequenza tende , per effetto della elevazione termica ,proporzionale alla temperatura tende a dissociare i nuclei . Questi essendo sostanzialmente le particelle (Muoni,neutrini neutroni protoni ect.ect) che originano i campi elettromagnetici {$\underline{k}$, $\underline{\hbar}$} del nucleo . Noi supponimo che la frequenza sia tale da rendere trascurabile la dissipazione e questa sia interamente termica R non ha memoria{$\underline{k}$, $\underline{\hbar}$} Si dice allora che R è ideale e privo di memoria Quindi se si chiude T_2(posizione on line) attraverso R fluisce una corrente $\boxed{i(t) = \frac{E_m}{R} \sin(\omega t)}$ (2) Si noti che la impedenza Z del circuito relativa al condensatore C e all'induttore L della (1) non dissipa energia dato che : $\boxed{W_L(t) + W_C(t) = 0}$ (3) per t ≥ 0 , se per t=0 è stato chiuso l'interruttore T_2, a seguire ..

It is useful to analyze the comportamnto of each element of analog circuits, both in static conditio invariant with respect to time, or variables in real fear. Spratutto, for the purpose of application and vary virtual time (s)

The fig2-TE17 shows the consequence of the separable components medians switches.

I-THE RESISTOR. This component is defined a a passive element of the **R.E** .And 'no memory electromagnetic. In the sense that all the energy supplied by the m.e.f. polarization impedance $\boxed{Z = R + j[L\omega - \frac{1}{\omega C}]}$ (1) is dissipated by Joule effect by R. At high frequencies the behavior of R is not likely to be associated with a dissipation electromagnetic field {$\underline{k}$, $\underline{\hbar}$ }. This is because the crystalline matter of resistors tends to convey the electrons of man - Tello nuclear So that growing in frequency tends, by effect of thermal elevation, proportional to-the temperature tends to dissociate the nuclei. These being essentially the particles (muons, neutrinos ect.ect neutrons protons) that originate electromagnetic fields {$\underline{k}$, $\underline{\hbar}$ } of the core. We supponimo that the frequency is such tha negligible dissipation and this is thfrmic R has no memory{$\underline{k}$, $\underline{\hbar}$)is then said R is ideal and no memory. So if you close T_2(position on line) in R,flows a current $\boxed{i(t) = (Em/R)\sin(\omega t)}$ (2) Note that the impedance Z of the circuit relative to the capacitor C and the inductor L of (1)not dissipation eneergi 0 given that: $\boxed{W_L(t) + W_C(t) = 0}$ (3) to t≥0, if at t = 0 the switch is closed T_2 , to follow ..

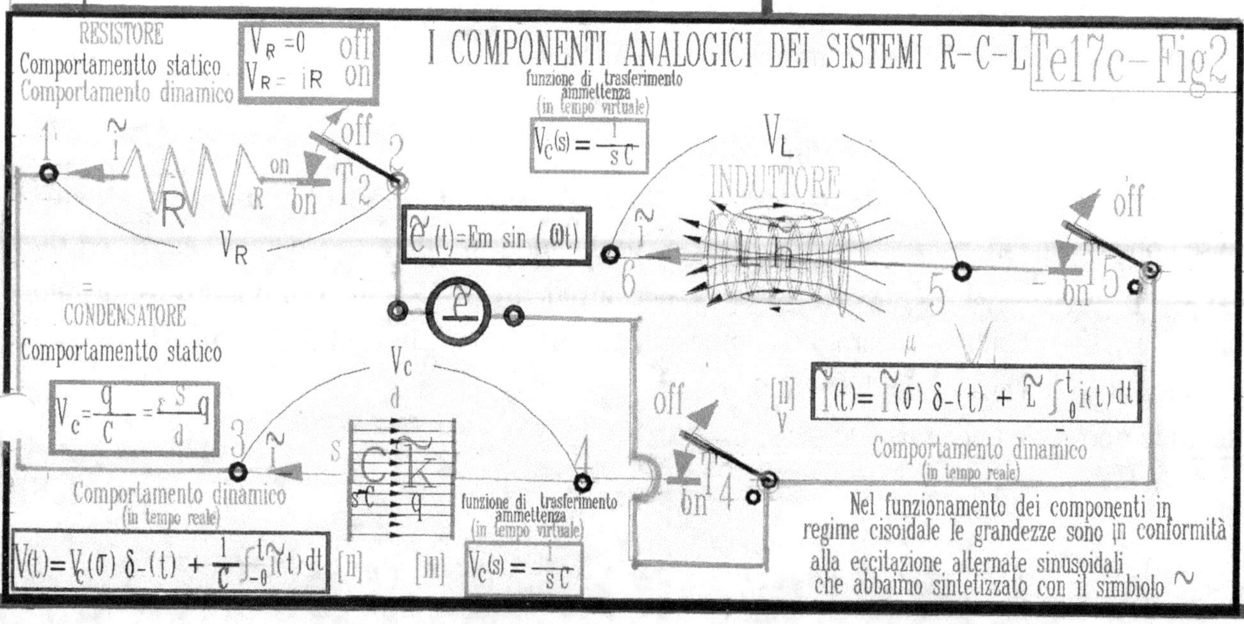

Consideriamo questi due elementi circuitali in regime permanente cisoidale, Fig20(a). Sfrondato da ogni complicazione è rappresentato da un sistema ricetrasmittente in bassa frequenza audio .

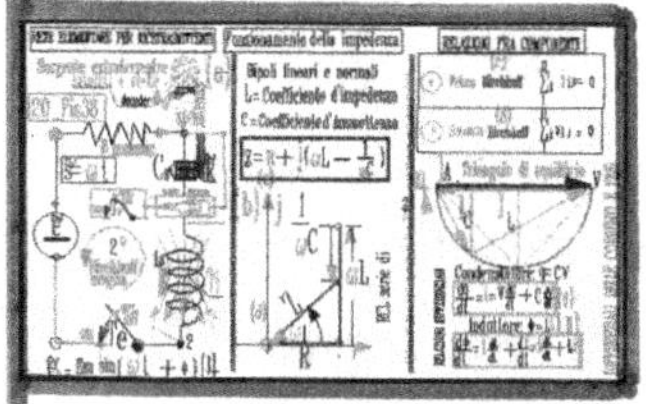

IL CONDENSATORE

Questo elemento forma pl parte integrante di tutti i dispositivi usati oggi per le telecomunicazioni , per trasmettere segnali di una banda estesa ,che va dai Giga Hertz(10^9 Hz)ai Mega Hertz(10^6Hz)

Tanto per fissare le idee.Questa è la questa frequenza di lavoro il Vostro P.C. Veniamo al dunque. La parola stessa di condensatore **C** suggerisce la idea di un contenitore,nel caso di cariche elettriche.

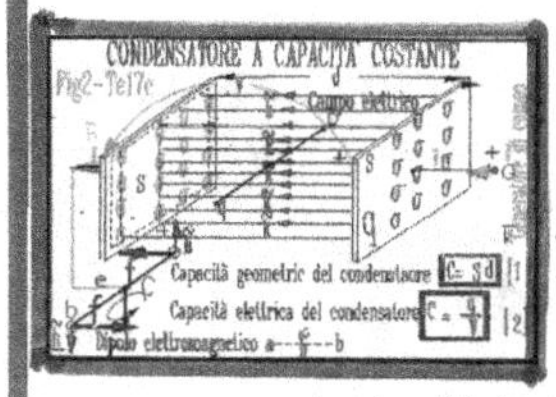

La tipologia costruttiva dei C è vasta Va dal **C** a lastre pianparallele, Fig 2 . Qui si sta faccendo riferimento a **C** per piccole capacità e frequenze e costanti formalizzato: $C=\varepsilon \frac{q}{V}$ (2) , ha la proprietà di intrappolare le cariche $q = \int_0^S \sigma\, ds$ (3) La distribuzione di cariche elemenatari σ crea un campo elettrico,$\bar{k}$ costante o tempo variante a seconda che la corrente i(t) sia questa o quella.Il fatto essenziale è che la corrente esterna a **C** crea un campo elettomagnetico $\bar{k},\bar{h}$ come mostra la barretta a-b associato. Il campo { $\bar{k},\bar{h}$ }è inseparabile alla sola condizione che la carica sia in movimento e non statica. Se ciò fosse il campo $\bar{k}\rightarrow$ diventa statico $\bar{k}$ Ma allora la corrente si spegne, interdetta da T_4 (off line)Fig 2 . Ma $\bar{k}$ è solo energia **q potenziale elettrica q che scaricata rigenera{$\bar{k},\bar{h}$ }**

Consider these two circuit elements in speed rate permanent cisoidae, Fig20 (a). Leafless from each complication is represented by a system trasmitter-riveiver in low frequency audio.

THE CONDENSER

This element pl form an integral part of all devices used today-you for telecommunications, signal of an extended band, ranging from in Giga Hertz (10^9 Hz) to Mega Hertz (10^6Hz)

Just to fix the idee. this is the working frequency aof your PC .This is to the point. The word itself of capacitor **C** suggests the idea of a container, in the case of electric charges.

The type of construction of the wide There is goes from **C** to plane-parallel plates, Fig 2.

Here are now being taken to **C** for small capacity and frequencies and somany formalized:

$C=\varepsilon \frac{q}{V}$ (2), has the property of trapping charges $q= \int_0^S \sigma\, ds$ (3) The fistibutios of charges elementary σ creates an electric $\bar{k}$ field, constant or time variant depending on the current i,(t) this is or quella.Il essential fact is that the external current to **C** creates a field as elettomagnetico shows $\bar{k},\bar{h}$ finger a-b associated. The field {$\bar{k},\bar{h}$ }} is inseparable on the sole condition that the charge is moving, rather than static. If this is the field $\bar{k}\rightarrow$ becomes static. But then the power turns off banned from T_4 (off line) Fig 2. But is just energy **q** and electric potential: q downloaded regenerates {$\bar{k},\bar{h}$}

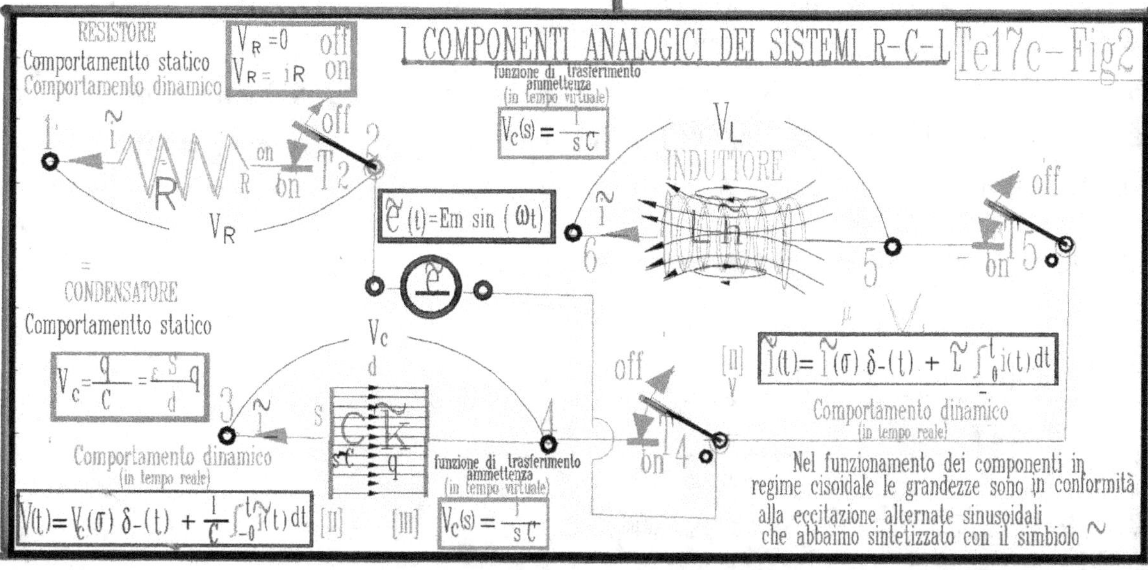

L' INDUTTORE

Ci sembra opportuno, prima di analizzare le caratteristiche circuitali della induttanza **L** e dare uno sguardo alle caratteritiche assiomatiche dei campi $\{\tilde{k},\tilde{h}\}$. Per giustificare i quali si stabilisce con l'esperienza la legge del legame materiale. postulata dalla induzione magnetica, $B=\mu\tilde{h}$ (1) unito allo spostamento elettrico ; $D=\varepsilon\,\tilde{k}$ (2) essendo μ,ε le rispettive permittività elettromagnetiche del mezzo , costanti se isotropo . Per questo la (1) e la (2) consentono di considerare il campo come equivalente ad una realtà fisica. Ma nella

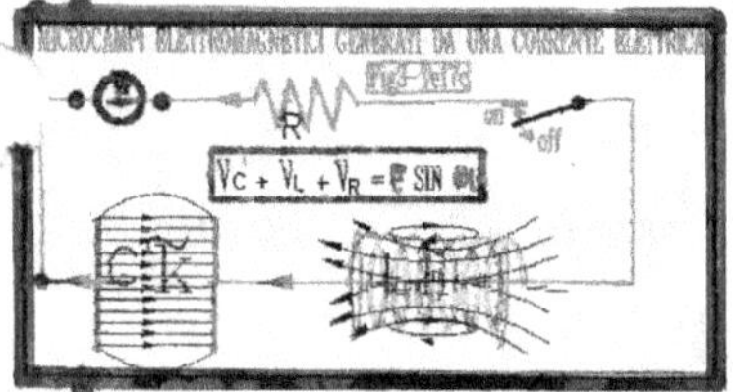

RCL se viene polarizzata dalla **f.e.m.**, Fig3, esiste una continuità materiale per cui non può esistere una campo dinamico attivato da una corrente elettrica senza che i campi (1) e (2) salvino la indissolubilià dei campi elettromagnetici dinamici ..Questo problema è stato risolto dal fisico **J.K.Mxwell** (1831-1892)

Creando un sistema di equazioni differenziali Preciamente con la eq,dif. : $\nabla \times \tilde{k} = -\frac{\partial B}{\partial t}$ (3) che lega il campo elettrico $\tilde{k}$ alla induzione magnetica B e il campo magnetico $\tilde{h}$ alla variazione tempo reale della densità elettrica associata alla corente di conduzione $\tilde{I}$ ed alla corrente impressa J_i definita dallo spostamento $\nabla \times \tilde{h} = \frac{\partial D}{\partial t} + \tilde{I} + J_i$ (4) Con riferimento al limitato supporto grafico si può identi ficcre i vettori $\tilde{k},\tilde{h}$ prodotti dalla induttanza L e da C, legati alla indu zione dalle permittività (1)e (2) Questi sono prodotti dalla corrente del circuito $\tilde{J}$ o,in alernativa da una f.e.m satellitaria impressa J_i che invia segnali digitali { a≡ 100000;1≡001} che il decoder converte in segnali Audio-Video

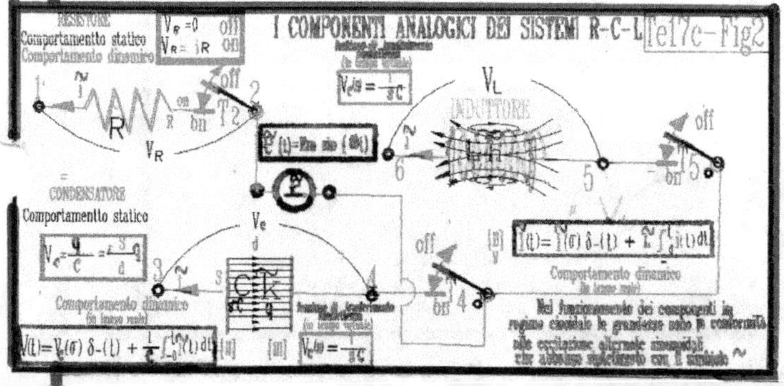

INDUCTOR

It seems appropriate, before analyzing the characteristics of the circuit inductance **L** and take a look at caratteritiche axiomatic field $\{\tilde{k},\tilde{h}\}$. To justify which are established with the experience the law of the bond material. postulated by magnetic induction,

$B=\mu\tilde{h}$ (1) combined with the electric displacement; $D=\varepsilon\,\tilde{k}$ (2) ,being μ,ε and the respective permittivity of the electromagnetic means, if isotropic constant. For this reason, (1) and (2) allow to consider the field as equivalent to a physical reality. But if in the RCL is polarized from the **e.m.f.**,, **Fig3**, there is continuity in the material for which it can not exist a dynamic field activated by an electric current without that the fields (1) and (2) to save hte indivisible of electromagnetic fields dynamic ..

This problem is been solved by physical J KMxwell (1831-1892) Inventor with the eq, dif. :

$$\nabla \times \tilde{k} = -\frac{\partial B}{\partial t} \quad (3)$$

that binds the electric field $\tilde{k}$ to the magnetic induction and magnetic field B to the change in real-time electrical density associated with courent of conduction $\tilde{I}$ and the impressed current J_i defined by the shift :

$$\nabla \times \tilde{h} = \frac{\partial D}{\partial t} + \tilde{I} + J_i \quad (4)$$

With reference to our limited graphics support can be identi-ficcre the vectors produced by the inductance L and the capacitor C connected to indution by the permittivity (1) and (2)

These are produced by the current of the circuit or, in alernativa by a fem satellitaria imprinted that sends the digital signals of the type {100000, 1001} that the decoder converts signals Audio-Video

RETE ELETTRICA A 7 MAGLIE

La **Fig10** rappresenta una **R.E**. di 7 maglie predisposta per la soluzione in notazione simbolica. Infatti le mutue impendenze di ammettenza, del

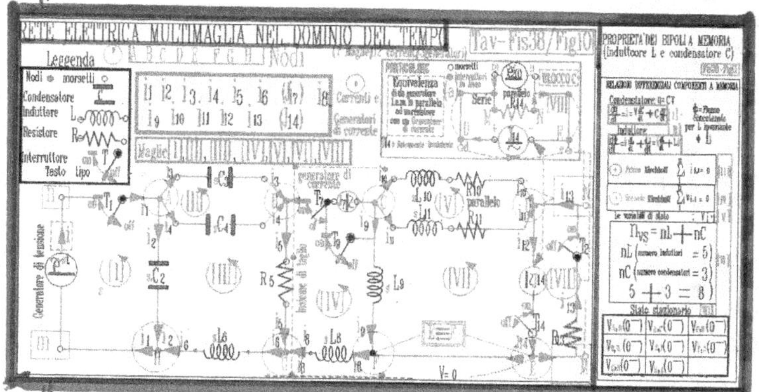

condensatore **C** e di Impedenza **L** in regime cisoidale costiutiscono la base del calcolo simbolico . A questo scopo si è indicato il condensatore come come una ammettenza complessa cisoidale s**C** Allo stesso modo la impedenza dell'induttore è sostituita dalla imepedenza s**L** .

Una soluzione di questa rete con i metodi delle equazioni differenziali in tempo reale richiederebbe la formulazione della equazione caratteristica e la soluzione della omogenea associata .

Per poi concludere con il metodo di **Lagrange** della variazione delle costanti. Si ricorre alla soluzione con il metodo di **Laplace** per le **R.E**. quando si presentano casi analoghi a quello della **Fig10** .Nonostante ciò il lavoro richiesto per la soluzione in correnti di nodo e tensioni dei lati risulta molto lungo. Si usa il metodo di Laplace in simbolica per il regime cisoidale operanto parzializzazioni della **R.E**. opportunamemte con insiemi di taglio e concludere con la legge della sovrapposizione degli effetti, siano questi ingressi e o uscite

in correnti e o tensioni

Siamo in tal modo giunti a concludere le modalità operative per la soluzione delle **R.E. in regime cisoidale** ($s=\sigma+j\omega$)

P.S. LAPLACE 1749-1827

POWER SUPPLY A 7 SHIRTS

The **Fig10** is a **R.E** 7 jerseys prepared for the solution in symbolic notation. In fact, the mutual impedances of admittance, the

The impedance of the capacitor **C** and impedadaance **L** under cisoidal cohorts have the basis of symbolic computation.

For this purpose, the capacitor is indicated as

as a complex admittance cisoidale s**C** Similarly,

the impedance of the inductor is replaced by impedenza s**L**.

A solution of this network with the methods of differential equations in real time would require the formulation of the characteristic equation and the solution of the associated homogeneous. And then conclude with the method of **Lagrange'**s variation of constants.

It makes use of the solution by the method of **Laplace** for RE when they show cases similar to that of Fig10.

Wich what the work required for the solution in currents and node voltages of the sides is very long. It uses the method of Laplace symbolic for the arming of the regime cisoidale operanto RE.opportunamemte with cut sets and conclude with the law of superposition, are these inputs and or outputs

voltages and or currents

We thus come to the conclusion the operating procedures for the solution of **R.E. in running cisoidal** ($s=\sigma+j\omega$)

RETE ELETTRICA A 7 MAGLIE

La **Fig10** rappresenta una **R.E**. di 7 maglie predisposta per la soluzione in notazione simbolica. Infatti le mutue impendenze di ammettenza, del

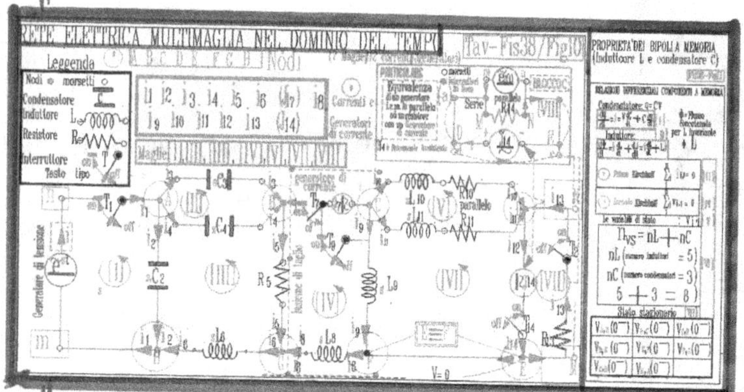

condensatore **C** e di Impedenza **L** in regime cisoidale costiutiscono la base del calcolo simbolico .
A questo scopo si è indicato il condensatore come come una ammettenza complessa cisoidale s**C**
Allo stesso modo la impedenza dell'induttore è sostituita dalla imepedenza s**L** .
Una soluzione di questa rete con i metodi delle equazioni differenziali in tempo reale richiederebbe la formulazione della equazione caratteristica e la soluzione della omogenea associata .
Per poi concludere con il metodo di **Lagrange** della variazione delle costanti. Si ricorre alla soluzione con il metodo di **Laplace** per le **R.E**. quando si presentano casi analoghi a quello della **Fig10**
.Nonostante ciò il lavoro richiesto per la soluzione in correnti di nodo e tensioni dei lati risulta molto lungo. Si usa il metodo di Laplace in simbolica per il regime cisoidale operanto parzializzazioni della **R.E**. opportunamemte con insiemi di taglio e concludere con la legge della sovrapposizione degli effetti, siano questi ingressi e o uscite
in correnti e o tensioni
Siamo in tal modo giunti a concludere le modalità operative per la soluzione delle **R.E. in regime cisoidale** ($s=\sigma+j\omega$)

POWER SUPPLY A 7 SHIRTS

The **Fig10** is a **R.E** 7 jerseys prepared for the solutio in symbolic notation. In fact, the mutual impedance of admittance, the

The impedance of the capacitor **C** and impedadaanc **L** under cisoidal cohorts have the basis of symbolic computation.

For this purpose, the capacitor is indicated as

as a complex admittance cisoidale s**C** Similarly,

the impedance of the inductor is replaced by impedenza s**L**.

A solution of this network with the methods of differential equations in real time would require the formulation of the characteristic equation and the solution of the associated homogeneous. And then conclude with the method of **Lagrange'**s variation of constants.

It makes use of the solution by the method of **Laplace** for RE when they show cases similar to that of Fig10.
Wich what the work required for the solution in currents and node voltages of the sides is very long. It uses the method of Laplace symbolic for the arming of the regime cisoidale operanto RE.opportunamemte with cut sets and conclude with the law of superposition, are these inputs and or outputs
voltages and or currents
We thus come to the conclusion the operating procedures for the solution of **R.E. in running cisoidal** ($s=\sigma+j\omega$)

Le auto e mutue impedenze $Z(j\omega)$ sono delle funzioni di trasferimento quando i segnali di corrente e/o tensione, sono sinusoidali e legano l'ingresso con l'uscita con eguale frequenza . La ammettenza $\dot{Y}(j\omega)$ è una funzione di trasferimento quando lega con la stessa frequenza i segnali di uscita ed ingreso fra la tensione e corrente dei singoli lati a componenti RCL

[P1] - i vantaggi del calcolo simbolico

1. Consente di sostituirevle equazioni differenziali equazioni in forma algebrica
2. Sostituire le d.d.p. con $j\omega C$ e $j\omega L$ dei componenti
3- Indicare Am con Ao e $j\omega = j2\pi f$ con la frequenza generlizzata $s=\sigma + j\omega$. Dopo queste premesse è facile intendere il significato della grandezza cisoidale $a(t)=A_o e^{\sigma t} \sin(\omega t+\alpha)$ (1)tradotta in formato simbolico nella corrispondente : $\dot{A}(t)=\dot{A}o\, e^{\sigma t}$ (2) , con $\dot{A}o=A_o e^{j\alpha}$ (3) Questa corrispondenza per la (1) è tale che a(t)è uguale al coefficiente dell'immaginario $a(t)=\operatorname{Im} \dot{A}o(t)$ (3') Il differenziale $dt=\frac{d^n a(t)}{dt^n}$ (4) diventa $D_o(t)=\frac{d^n A_o(t)}{dt^n} = s^n \dot{A}_o(t)$ (5)

Le equivalenze polinomiali di uscita $y(t) \rightarrow$ ingresso u(t)in simbolica risultano definita dalla equivalenza :

$$\dot{Y}_o(t) \sum_{i=0}^{n} a_i s^i = \dot{U}_o(t) \sum_{i=0}^{m} b_i s^i \quad (6)$$

,dalla quale si ottiene la uscita simbolica separando $\dot{Y}_o(t)$; Nel fatto risulta:

$$\dot{Y}_o(t)=\dot{U}_o(t) \left[\left(\sum_{i=0}^{m} b_i s^i : \sum_{i=0}^{n} a_i s^i \right) \right] = \dot{U}_o(t)\ W(s)= \dot{U}_o(t) \frac{b_m s^m + b_{m-1} . s^{m-2} + \ldots + b_1 s + b_o}{a_n s^n + a_{n-1} . s^{n-1} + \ldots + a_1 s + a_o} \quad (7)$$

Si può dire che la uscita in simbolica $\dot{Y}_o(t)$ è il prodotto dell'ingresso $\dot{U}_o(t)$ per la f di trasferimento

$$W(s) = \frac{b_m s^m + b_{m-1} . s^{m-2} + \ldots + b_1 s + b_o}{a_n s^n + a_{n-1} . s^{n-1} + \ldots + a_1 s + a_o} = \frac{\dot{V}}{\dot{I}} \quad (8)$$

La funzione di traferimento consente di risolvere i problemi delle reti elettriche dei segnali cisoidali di ingresso ed uscita legati alle tensioni e le correnti dei singoli lati e delle maglie della rete in regime forzato

The self and mutual impedancesZ (jw) are func-tions of transfer when the signals of current and / or voltage are sinusoidal and bind the input to the output with equal frequency. The admittance (jw) is a transfer function when it binds with the same frequency output signals and ingreso between the voltsge and current of the single-sided components RCL

P1 - the advantages of symbolic computation

1. Lets sostituirevle differential equations equations in algebraic form
2. Replace d.d.p. with $j\omega C$and $j\omega L$ component
3 - To state Am with Ao and $j\omega = j2\pi f$ with the frequency to generalized $s = \sigma + j\omega$. After these facult to understand the meaning of greatness cisoidale $a(t) = A_o e^{\sigma t} \sin(\omega t+\alpha)$ (1) translated into symbolic in the corresponding format $\dot{A}(t)=\dot{A}o\, e^{\sigma t}$ (2), with $\dot{A}o=A_o e^{j\alpha}$ (3) This correspondence for the (1) is such that a (t) is equal to the imaginary coefficient of $a(t) = \operatorname{Im} \dot{A}o(t)$ (3'). The differential $dt = dt=\frac{d^n a(t)}{dt^n}$ (4) becomes $D_o(t)=\frac{d^n A_o(t)}{dt^n} = s^n \dot{A}_o(t)$ (5)

The equivalences polynomial output y (t) input u (t) are defined by the symbolic equivalence:

$$\dot{Y}_o(t) \sum_{i=0}^{n} a_i s^i = \dot{U}_o(t) \sum_{i=0}^{m} b_i s^i \quad (6)$$

from which is obtained the output symbolic separating; In fact results:

$$\dot{Y}_o(t)=\dot{U}_o(t) \left[\left(\sum_{i=0}^{m} b_i s^i : \sum_{i=0}^{n} a_i s^i \right) \right] = \dot{U}_o(t)\ W(s)= \dot{U}_o(t) \frac{b_m s^m + b_{m-1} . s^{m-2} + \ldots + b_1 s + b_o}{a_n s^n + a_{n-1} . s^{n-1} + \ldots + a_1 s + a_o} \quad (7)$$

It can be said that the symbolic$\dot{Y}_o(t)$ output is the product of the input $\dot{U}_o(t)$ for the transfer f

$$W(s) = \frac{b_m s^m + b_{m-1} . s^{m-2} + \ldots + b_1 s + b_o}{a_n s^n + a_{n-1} . s^{n-1} + \ldots + a_1 s + a_o} = \frac{\dot{V}}{\dot{I}} \quad (8)$$

The transfer must function allows to solve the problems of electrical networks of signals cisoidali input and output related to the voltages and currents of the individual sides of the meshes **of the network**

C.A. COULOMB

(1736 - 1808)

Le reti elettriche costituite da bipoli ideali di mutuo induttori schema di **Fig5**

[a] Un altro schema equivalente **Fig2-Inda2**

Sono deducibili le relazioni tension-correnti fra i ngresso ed uscita del trasformatore

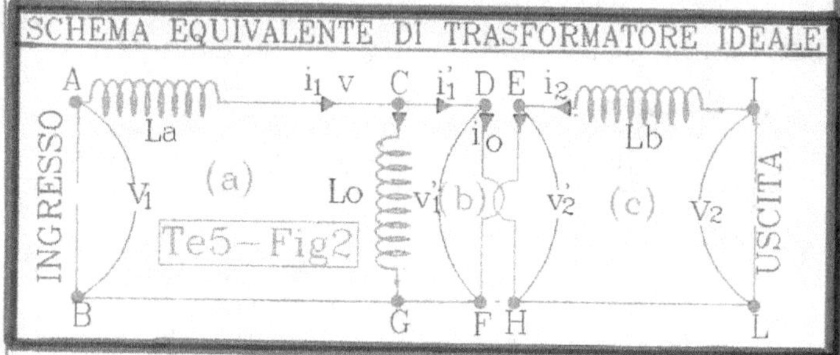

[b] Le relazioni correnti e tensioni

Precisamente

$i_1=i'_1+i_o$ (1) , $v'_1=n.v'_2$ (2) , $i'_1=-\frac{1}{n}i_2$ (3)

$v'_1=L_o\frac{di_o}{dt}$ (4) , $V_1=L_a\frac{di_1}{dt}+V'_1$ (5) , $V_2=L_b\frac{di_2}{dt}\ v'_2$ (6)

Dalle (1),(3),(4) si ottiene : $V'1=Lo[\frac{di_1}{dt}-\frac{di'_1}{dt}]=$ $=Lo[\frac{di_1}{dt}+\frac{1}{n}\frac{di_2}{dt}]$ (7) . Dalla (5) e (2) si trovano le : $v'_2=\frac{L_o}{n}[\frac{di_1}{dt}+\frac{1}{n}\frac{di_1}{dt}]$ (8) . Sostituendo nella (7) , $V_1=(La+Lo)\frac{di_1}{dt}+\frac{L_o}{n}\frac{di_2}{dt}$ (9) . Allora confrontando la (6) con la (9)si può considerare l'induttore doppio , nicchia (a)+ (b) fra loro equivalenti e quindi se isi pone $La+Lo=L_1$ (10) , $\frac{L_o}{n}=M$ (11). Anche: $L_b+\frac{L_o}{n^2}=L_2$ (12) . Le (19(11)(12) possono essere soddisfatte attribuendo un valore arbitrario ad uno dei quattro parametri L_a,L_b,L_o, n . In particolare......

Electrical networks consist of dipoles ideals of mutual inductors scheme **Fig5**

[a] Another equivalent diagram **Fig2-Inda2**

Are deductible tension-current relations between his entry and exit of the transformer

[b] The relations currents and voltages

Precisely :

$i_1=i'_1+i_o$ (1) , $v'_1=n.v'_2$ (2) , $i'_1=-\frac{1}{n}i_2$ (3)

$v'_1=L_o\frac{di_o}{dt}$ (4) , $V_1=L_a\frac{di_1}{dt}+V'_1$ (5) , $V_2=L_b\frac{di_2}{dt}\ v'_2$ (6)

From (1), (3), (4) is obtained:

The: $V'1=Lo[\frac{di_1}{dt}-\frac{di'_1}{dt}]=Lo[\frac{di_1}{dt}+\frac{1}{n}\frac{di_2}{dt}]$ (7)

From (5) and (2) are located: $v'2=\frac{L_o}{n}[\frac{di_1}{dt}+\frac{1}{n}\frac{di_1}{dt}]$ (8)

Substituting into (7) : $V_1=(La+Lo)\frac{di_1}{dt}+\frac{L_o}{n}\frac{di_2}{dt}$ (9) .

then by comparing the(6) with (9) can be considered the inductor double, niche (a) + (b) between their equivalents, and then if isi arises the :

$La+Lo=L_1$ (10) , $\frac{L_o}{n}=M$ (11) .

also: $L_b+\frac{L_o}{n^2}=L_2$ (12) . The (19 ..(11) (12) may be fulfilled by assigning an arbitrary value to one udei four parameters L_a, L_b, L_o, n. In particular ...

O.Heaviside (1850-1925)

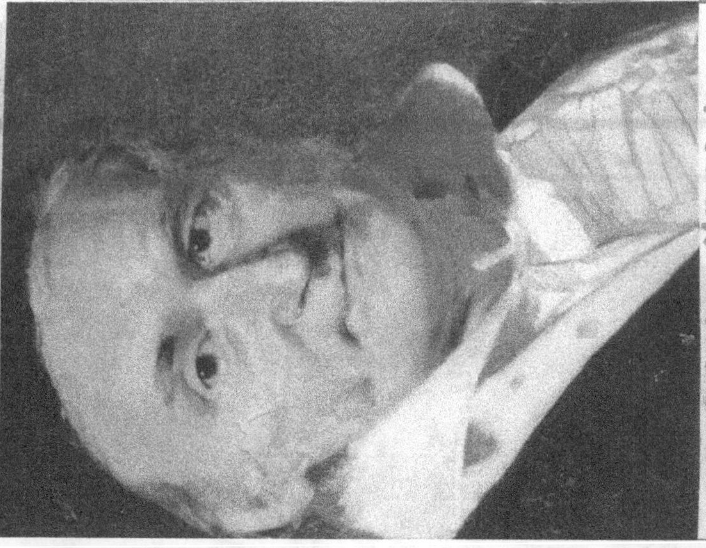

C.F.GAUSS (1777-1855)

[a]cariche e correnti elettriche Per definizione si scrive : $i(t)=\lim_{\Delta t\to 0}\frac{\Delta q}{dt}$ (1) , **Δq**(variazione istantanea della carica q che attraversa la sezione di un conduttore nella unità di tempo arbitraria) Nel S.I. si misura in A(Ampere)=1(Coulomb) s^{-1} (**Ap-Z**/Tav-II) In un buon conduttore o su un(s.c) **semiconduttore** drogato , la carica q intesa come flusso di elettroni attraversa un bipolo , se la d.d.p. (polarizzazione) sposta gli elettroni in **banda conduzione** , Fig53-Tee28 . E nel caso di un(s.c drogato),**Fig 30-Micro1** dai Chip della micro elettronica .[Nella (**Ap-Z**/Ib-Tab B) , le cortecce:K-L-M-....Q, e le sotto-cortecce s-p-t del mantello elettronico.Il nucleo del Silicio(s.c) ha 14 elettroni ed il Germanio (s.c) ne possiede 32]

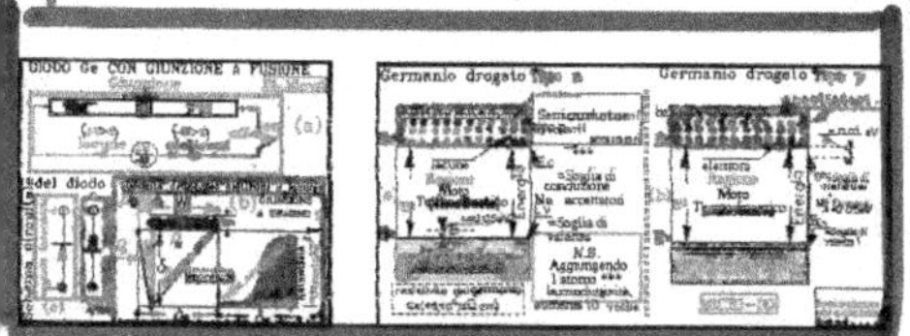

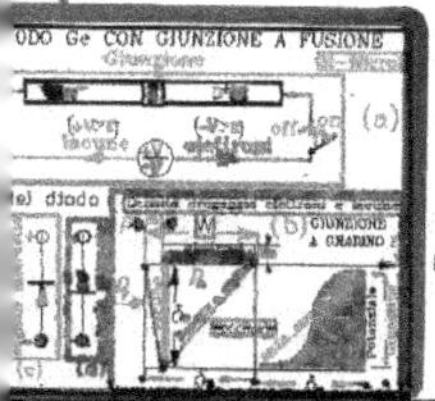

Questa escursione fra i s.c. tiene conto del fatto che allo sviluppo delle reti elettriche classiche si sono aggiunte i Chip (reti miniaturizzate e stampate) Queste hanno consentito lo sviluppo delle reti di trasmissione dei segnali in frequenza digitale. Da cui nascono i telefonini . Inoltre una varietà di Chip, dagli ultrasuoni alla robotizzazione degli autoveicoli Per finire alla fisica planetaria del triofo della fisica dei grandi teorizzatori, da **Plank** ad **Einstein** .Ma pure dell'Homo Faber. Una catena di satelliti artificiali(progenitore **Von Braun**). Sfruttando la dimanica di **Newton** (**Kepler docet**) sono approdati in orbita geostazionaria un sistema planetario di satelliti, che consentono delle trasmissioni Tv digitali in quanto dotati di **LNB**,Fig2 emettono segnali in alta frequenza(video), e bassa audio.I segnali digitali emessi sono convertiti poi dai decoder terrestri su aree del geoide terrestre

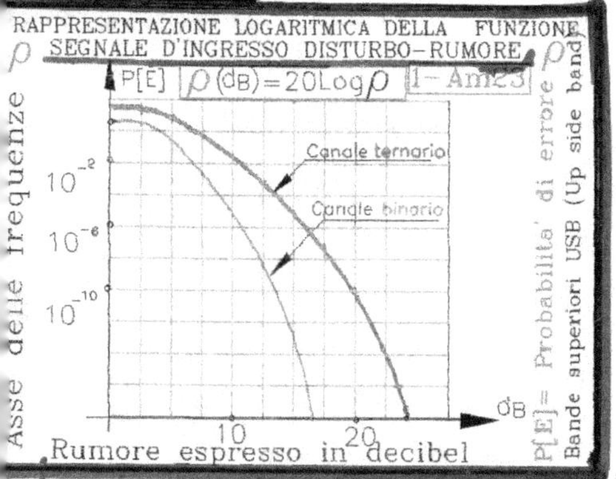

[a] electrical charges and currents, by definition you write:$i(t)=\lim_{\Delta t\to 0}\frac{\Delta q}{dt}$ (1) , **Δq** (Instantaneous change in the charge q passing through the section of a conductor in arbitrary units of time)In SI is measured in A(Ampere)= 1 (Coulomb) s^{-1} (**Ap-Z**/Tav-II) In a good conductor or of a (sc)-doped semiconductor, the charge q understood as the flow of electrons through a circuit element, if the ddp (Polarization) moves the electrons in **conduction band**, **Fig53-Tee28.** And in the case of a (s.c doped), **Fig 30-micro1** by Chip of micro electronics. [In (Ap-Z/Ib-Tab B), barks: KLM-.... Q, and the under-bark spt coat elettronico.Il core of silicon (s.c) has 14 electrons and Germanium (s.c) owns 32] This excursion between the sc takes into account the fact that the development of electricity networks were added the classic Chip (networks miniaturization authorized and printed) These have allowed the development of the transmission of signals in frequency gitale. From where I was born - no mobile phones. Furthermore a variety of chips, from the ultrasonic-ni robotization of motor vehicles To finish the physics of planetary triofo physics major theo-timers, from **Plank** to **Einstein.** But even Homo Faber. A chain of artificial satellites (progenitor **Von Braun**). Leveraging dimanica **Newton** (**Kepler** docet) have come into geostationary orbit a planetary system of satellites, which enable digital TV broadcasts in quantodotati **LNB**, Fig2 emit high-frequency signals (video), and low audio. The digital signals emitted are then converted by the decoder of terrestrial areas of the geoid

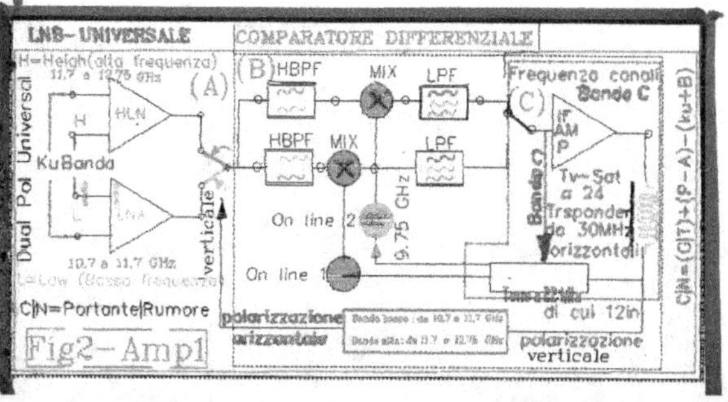

Fig2-Amp1

LA CORRENTE NEI CONDUTTORI E NEI SEMICONDUTTORI

(traduzione leggenda frafici)

apg-68a

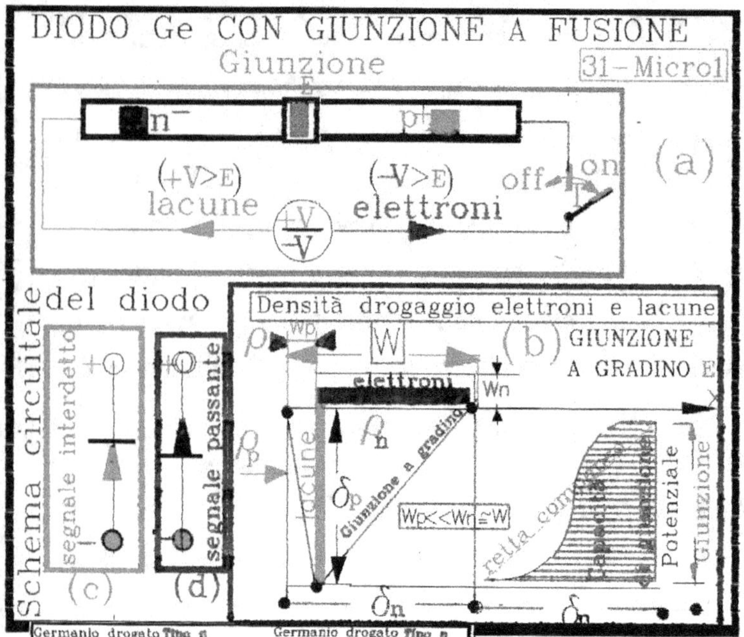

DIODO Ge CON GIUNZIONE A FUSIONE

n(elettroni)

p(lacune)

DIODO Ge GIUN-TIO[N]
WITH A MERGER

n (electrons)

p (gaps)

Schema circuitale del diodo

Densità del drogaggio

Giunzione a gradino di elettroni

Circuit diagram of the dio[de]

Density of doping

Junction step electron

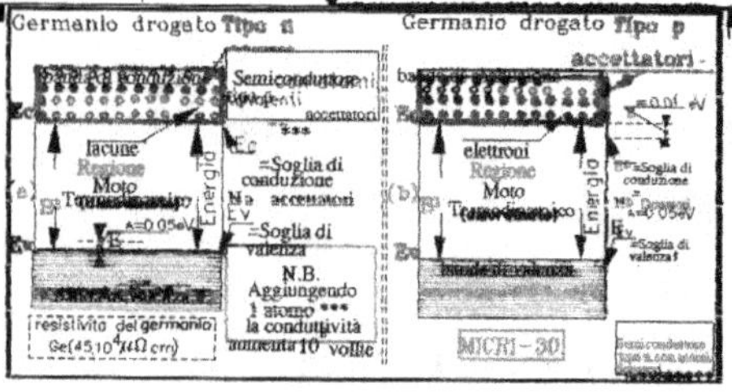

DIODO tipo **n-p**
accettatori**(elettroni in banda** di conduzione)

Regione proibita

elettroni di valenza appartenemti al nucleo

DIODE type **n-p**
acceptors (**electrons in ba[nd]** conduction)

forbidden region

valence electrons to th[e] nucleus appartenemti

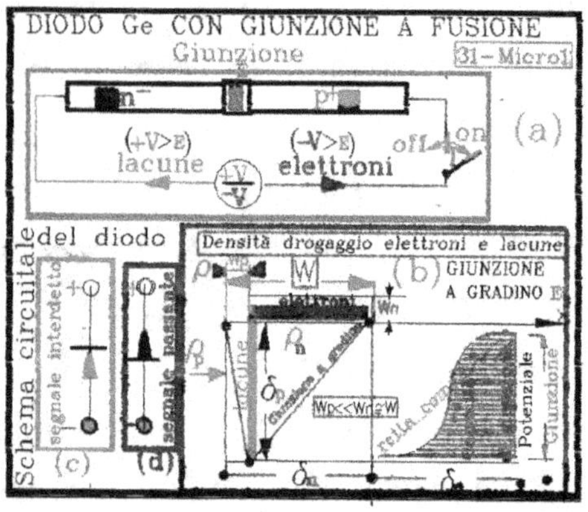

DIODO Ge CON GIUNZIONE A FUSIONE

lacune elettroni

Ge JUNCTION DIOD[E]
WITH A MERGER

gaps electrons

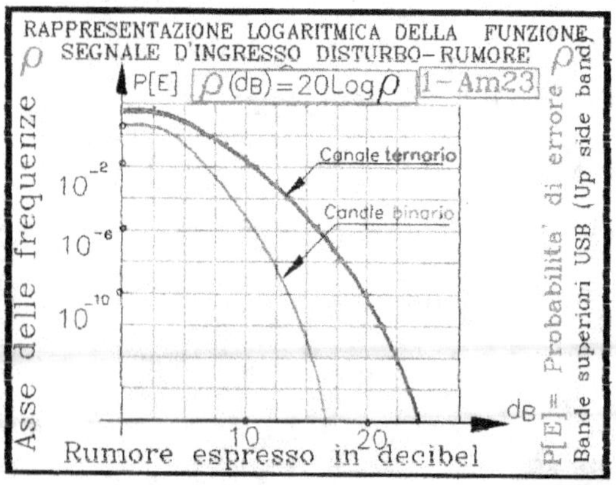

Segnle di diisturbo dei segnali (asse frequenze- assedecibel)

Signal of diisturbo signal (frequency-axis assedecib[el])

Comparatore differenziale per alte frequenze(HLN da 11,7 a 12,76 GHz)

Differential comparator [f] high frequencies (HLN 11.[7] 12.76 GHz)

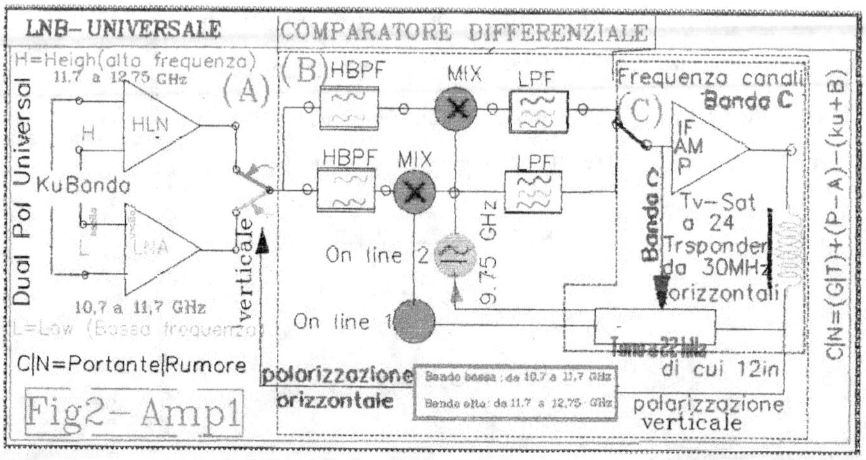

LA CORRENTE NEI CONDUTTORI E NEI SEMICONDUTTORI

(traduzione leggenda ʃrafici)

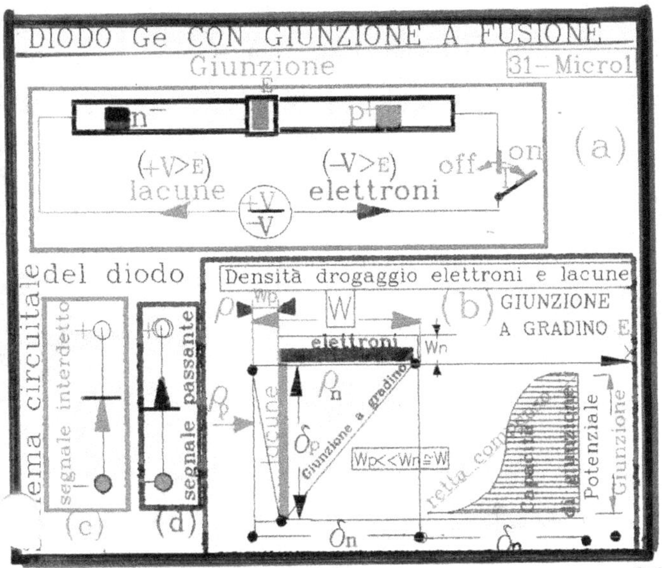

DIODO Ge CON GIUNZIONE A FUSIONE
n(elettroni)
p(lacune)

DIODO Ge GIUN-TION WITH A MERGER
n (electrons)
p (gaps)

Schema circuitale del diodo
Densità del drogaggio
Giunzione a gradino di elettroni

Circuit diagram of the diode

Density of doping
Junction step electron

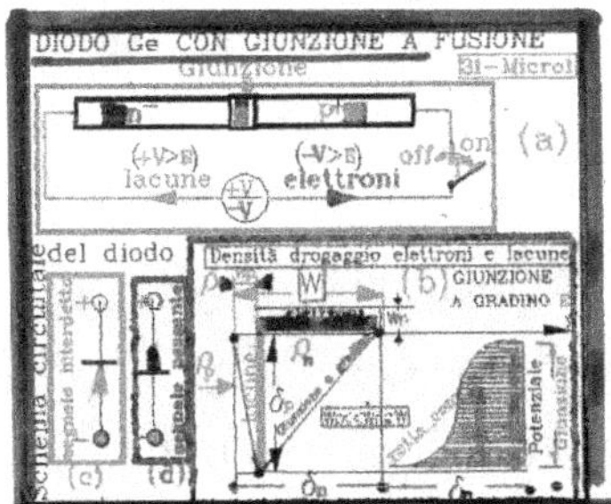

DIODO tipo n-p
accettatori(**elettroni in banda** di conduzione)
Regione proibita
elettroni di valenza appartenemti al nucleo

DIODE type n-p
acceptors (**electrons in band** conduction)

forbidden region
valence electrons to the nucleus appartenemti

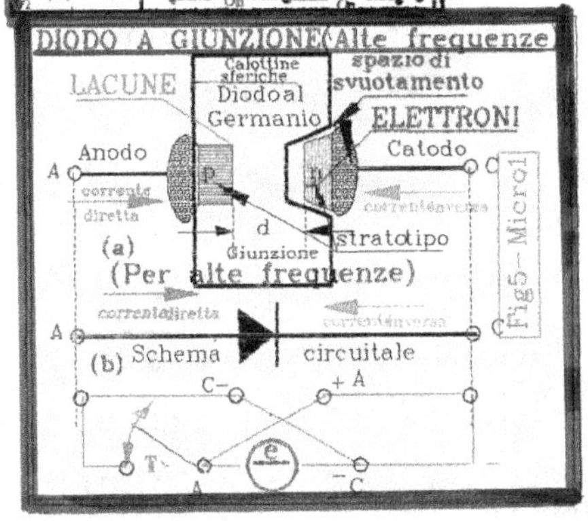

DIODO Ge CON GIUNZIONE A FUSIONE
lacune **elettroni**

Ge JUNCTION DIODE WITH A MERGER
gaps **electrons**

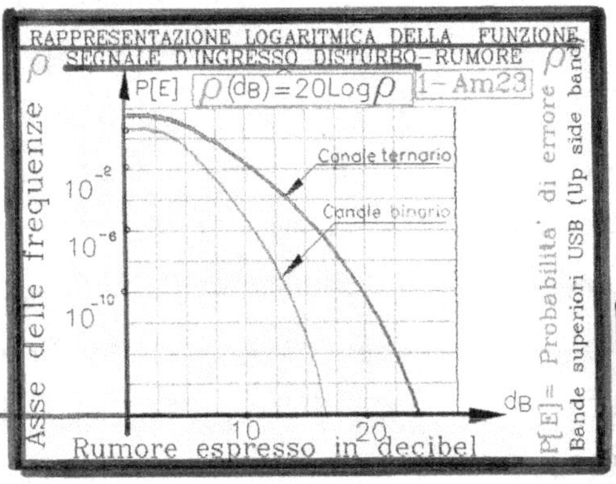

Segnle di diisturbo dei segnali (asse frequenze- assedecibel)

Signal of diisturbo signal' s (frequency-axis assedecibel)

Comparatore differenziale per alte frequenze(HLN da 11,7 a 12,76 GHz)

Differential comparator for high frequencies (HLN 11.7 to 12.76 GHz)

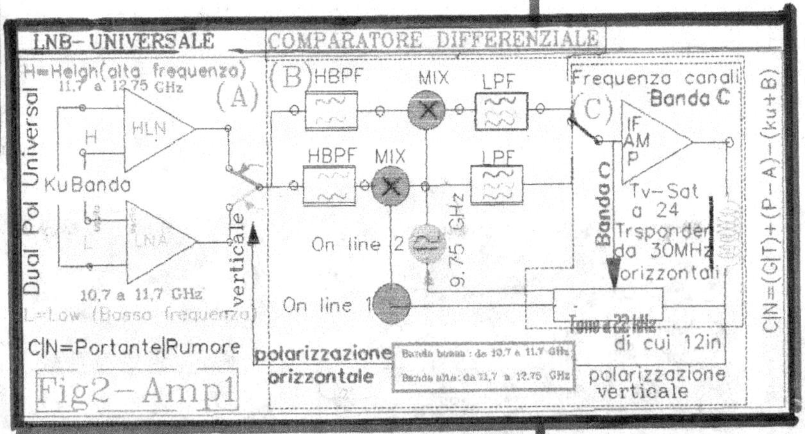

Le auto e mutuo impendenze sono funzioni di trasferimento quando l'ingresso sia del tipo sinusoidale . In tal caso le reti elettriche possono essere risolte con

IL METODO SIMBOLICO .

[a] - Definizione della componentistica di una **R.E.** (Rete Elettrica) . In generale si fa riferimento alla uscita k-esima del componente generico ,k=1,2,..,n . Il numero dei bipoli della rete , distinti con i grafi : l=ngv+ngi+......+ nL +nC+nR (1) rispettivamente generatori di corrente e o tensione induttanze **L** e conduttanze **C** .Resistori **R**. Va poi tenuto conto che la corrente e o tensione va ricercata singolarmente bipolo per bipolo e in tal caso è inutile la distinzione.

[b] Vantaggi del calcolo simbolico

Abbiamo visto a **pg-51** la soluzione in tempo reale della corrente alternata dell'oscillatore con il metodo delle equazioni differenziali dei componenti R-C-L . risolta con il metodo di Lagrange: $y(t)=y_o(t)+y_p(t)$ (2) Osservate dal lato fisico consentono di evidenziare le caratteristiche in versione operativa sono, specie se la R.E.è multimaglia, molto laboriosa da risolvere.

Il metodo simbolico di fatto presenta i vantaggi :

1° # Di scrivere le equazioni di maglia del secondo principio di **Kirchhoff** in forma algebrica invece che in forma differenziale #

2° # Di esprimere correnti in uscita dei singoli lati di ogni maglia con il primo principio di Kircchoff #

LE RELAZIONI ALGEBRICHE SIMBOLICHE

Allo scopo. Detta Ao(t) la grandezza cisoidale in tempo reale si pone A(jω) in simbolica. In tal modo :

3° # Le auto e mutue impedenze Ż(jω) sono funzioni di trasferimento Ẇ(jω) quando l'ingresso sia una corrente sinusoidale e l'uscita una tensione sinusoidale di uguale frequenza #

4° #Le auto e mutue ammettenzeẎ(jω) sono funzioni di trasferimento per tensioni(ingresso) e correnti(uscita) #

GENERALIZZAZIONE DEL CALCOLO SIMBOLICO

Se al posto di jω si pone s=σ+jω si può scrivere la uscita $a(t)=Ao\, e^{\sigma t} \sin(\omega t+ \alpha)$ (2) in tempo reale

The self and mutual impedances are functions of transfer when the input is of type sine wave. In this case, the electric grids can be solved

with symbolic methods.

[a]-Definition of the components of a **RE** (Power Grid). In general it refers to the output of the k-th generic component, k = 1,2, .., n. The number of dipoles in the network, with separate graphs: l =ngv+ngi + ... + nL + nC + nR (1), respectively power generators and or voltage inductances **L** and **C** conductances. Resistors **R**. It should also be taken into account that the current and or voltage must be sought individually the bipolar-to-port and in that case the distinction is unnecessary.

[b] Advantages of symbolic computation

We have seen in **pg-51** solution in real time of the alternating current of the oscillator with the method of differential equations of the components RCL, resolved by the method of Lagrange: $y(t)=y_o(t)+y_p(t)$ (2) .Observe from the physical side allow to highlight the features in operational version are, especially if the REE multimaglia,very laborious to solve. The symbolic method that has the advantages of: 1° # To write the loop equations of the second law of **Kirchhoff** in algebraic form rather than as a second differential #

2°# To express Outbox currents of the individual sides of each mesh with the first principle of **Kir.**#

ALGEBRAIC RELATIONS SYMBOLIC

The purpose. This Ao (t) is the cisoidale in tem-bit real arises A (jω) in symbolic. In this way:

3° # The third self and mutual impedances are Ż(jω) transfer functions when the input is a sinusoidal current and the output voltage sine wave of the same frequency. 4° # The self and mutual admittances are transfer functions for voltages and current #

GENERALIZATION OF SYMBOLIC COMPUTATION

If instead of jw arises s = σ + jω we can write: $a(t) = Ao\ e^{\sigma t} \sin(\omega t + \alpha)$ (2) in times real

1667 – 1754

A. De Moivre

(1821-1894)

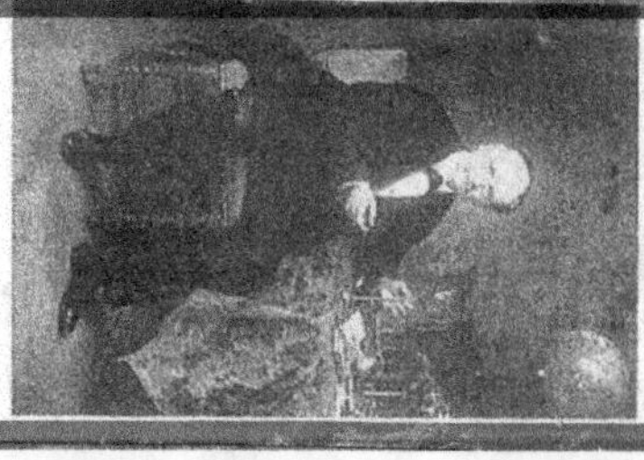

Helmholtz

Esaminiamo in dettaglio le conclusioni precedeenti

3° # Le auto e mutue impedenze $\dot{Z}(j\omega)$ sono funzioni di trasferimento $\dot{W}(j\omega)$ quando ingreso ed uscita sono sinusoidali e seguale frequenza #

4° #Le auto e mutue ammettenze $\dot{Y}(j\omega)$ sono funzioni di trasferimento da tensioni(ingresso) a correnti(uscita) sinusoidali di eguale frequenza #

LA GENERALIZZAZIONI SIMBOLICHE

[a] Le relazioni da reali a simboliche

1°- La corrispondenza reale - simbolica della a(t) :

$$a(t)=A_o e^{\sigma t} \sin(\omega t+\alpha) \rightarrow \dot{A}_o(t) e^{st} \quad (1)$$

anche → $\dot{A}_o(t) e^{st} = A_o e^{j\varphi} e^{st} = A_o e^{\sigma t+j(\omega t+\varphi)}$ (2)

2°-La corrispondenza è tale che a(t) risulta uguale al coefficiente dell'im. $Im[\dot{A}_o(t)] = A_o e^{st}$ (3)

3°- Le derivate si scrivono allora in simbolica :

$$\rightarrow \quad \dot{D}_o(t) = \frac{d^n \dot{A}_o(t)}{dt^n} = s^n \dot{A}_o(t) \quad (4)$$

[b] La uscita $\dot{Y}_o(t)$ è legata all'ingresso $\dot{U}_o(t)$ dalla corrispondenza : $\dot{Y}_o(t) = \dot{U}_o(t) . \dot{W}(s)$ (5) Fig2

$$\text{Con: } \dot{W}(s) = \frac{b_n s^n + b_{n-1} s^{n-1} + ... + b_1 + b_o}{a_n s^n + a_{n-1} s^{n-1} + ... + a_1 + a_o} \quad (7)$$

che prende il nome di Funzione di Trasferimento **F.d.T.** in regime cisoidale .Per una definita frequenza cisoidale s $\dot{W}(s)$ è un numero **complesso costante**

[c] Una forma operativa della F.d.T.

Si pone la **F.d.T**in forma: $\dot{W}(s) = W(s) e^{j\varphi}$ (8) .Da cui

$$\dot{Y}_o(t) = \dot{Y}_{oe}^{st} = \dot{W}(s) \dot{U}_o = W(s) U_o e^{j(\phi+\alpha)} e^{st} = Y_o e^{j(\phi+\alpha)} e^{st} \quad (9)$$

Anche $Y_o = W(s) U_o$ (10) .Infine la partticolare forma.

$Yp(t) = Y_o e^{\sigma t} \sin(\omega t+\alpha+\varphi)$ (11) per ingressi cisoidali

Abbiamo in sintesi la formulistica necessaria per poter risolvere le reti elettriche a più maglie per mezzo di equazioni algebriche invece di equazioni differenziali

Let us examine in detail the findings precedeenti

3° # The third self and mutual impedances $\dot{Z}(j\omega)$ transfer functions $\dot{W}(j\omega)$ when ingreso and output are equal frequency sinusoidal and #

4° # The self and mutual admittances are transfer functions from voltages (input) to current (output) of equal frequency sinusoidal #

LA SYMBOLIC GENERALIZATIONS

[a] The reports from real symbolic

1°-The royal correspondence - symbolic of a(t):

$$a(t)=A_o e^{\sigma t} \sin(\omega t+\alpha) \rightarrow \dot{A}_o(t) e^{st} \quad (1)$$

also→ $\dot{A}_o(t) e^{st} = A_o e^{j\varphi} e^{st} = A_o e^{\sigma t+j(\omega t+\varphi)}$ (2)

2°-The correspondence is such that a (t) is equal to the coefficient of im. : $Im[\dot{A}_o(t)] = A_o e^{st}$ (3)

3°- The third-derivative are written in symbolic

$$\rightarrow \quad \dot{D}_o(t) = \frac{d^n \dot{A}_o(t)}{dt^n} = s^n \dot{A}_o(t) \quad (4)$$

[b] The output $\dot{Y}_o(t)$ is related to entry $\dot{U}_o(t)$ in the correspondence: $\dot{Y}_o(t) = \dot{U}_o(t) . \dot{W}(s)$ (5) Fig2

$$\text{With } \dot{W}(s) = \frac{b_n s^n + b_{n-1} s^{n-1} + ... + b_1 + b_o}{a_n s^n + a_{n-1} s^{n-1} + ... + a_1 + a_o} \quad (7)$$

which is called the Transfer Function **F.d.T.** cisoidale.For under **a defined frequency** cisoidale s,$\dot{W}(s)$ is a **complex** number constant

[c] A form of operational F.d.T. It raises the **F.d.T**.in form: $\dot{W}(s) = W(s) e^{j\varphi}$ (8). from which

$$\dot{Y}_o(t) = \dot{Y}_{oe}^{st} = \dot{W}(s) \dot{U}_o = W(s) U_o e^{j(\phi+\alpha)} e^{st} = Y_o e^{j(\phi+\alpha)} e^{st} \quad 9$$

Also: $Yo = W(s) Uo$ (10) and the particular form

$Yp(t) = Y_o e^{\sigma t} \sin(\omega t+\alpha+\varphi)$ (11) for a summary of its inputs cisoidali We formulistica required to solve the electricity network to multiple meshes using algebraic and nont equation diffrential

NGRESSO– USCITE E LA FUNZIONE DI TRASFERIMENTO

Am13–Fig2

$$\dot{Y}_o(t) = \dot{U}_o(t) \frac{\sum_{i=0}^{m} b_i s}{\sum_{i=0}^{n} a_i s} = \dot{U}_o(t) \frac{b_m \cdot s^m + b_{m-1} \cdot s^{m-1} + \cdots + b_1 \cdot s + b_o}{a_n \cdot s^n + a_{n-1} \cdot s^{n-1} + \cdots + a_1 \cdot s + a_o}$$

consideriamo la corripondenza :

$Y_o(t)=W(s)U_o(t)$ (1) $Y_p(t)=Y_oe^{\sigma t}\sin(\omega t+\alpha+\varphi)$ (2)

La (2) $Y_p(t)$ è un integrale particolare se si considera come funzione di ingresso : $u(t)=U_oe^{\sigma t}\sin(\omega t+\alpha)$ (3) Si può ottenere la (2) con il metodo simbolico un integrale particolare in luogo della equazione differenzuiale di **pg 45** [I] e **[II]** . Quando l'ingresso è del tipo (2) di frequenza generalizzata $s=\sigma+ j\omega$ (a) con fase a l'argomento φ di W(s) . Nel calcolo si indica con $Z(s)=\omega L$ e $Y(s)=\omega C$ rispettivamente la impedenza e la ammettenza simbolica, **Fig3**

LA FUNZIONE DI TRASFERIMENTO Am13–Fig3

$$\dot W(s)=\frac{b_m\cdot s^m+b_{m-1}\cdot s^{m-1}+\dots+b_1\cdot s+b_o}{a_n\cdot s^n+a_{n-1}\cdot s^{n-1}+\dots+a_1\cdot s+a_o}$$

La conoscenza della **F.d.T** $\dot W(s)$ è suffiente per determinare la $\phi(t)$, lo sfasamento , che emerge quando si ponga: $\dot W(s)=W(s)\,e^{j\phi t}$ (4), essendo W(s) il modulo di W(s).L'integrale particolare (2) in notazione simbolica presenta le grandezze di ingresso ed uscita ,in **Fig17** la f.e.m. $\tilde e(t)=Em\sin(\omega t+\alpha)$ rappresenta l'ingresso u(t) in tempo reale $\dot U(s)=Uo\,e^{st}$ (5)

[a] L'integrale di P.S.Lapalace(1749-1827)

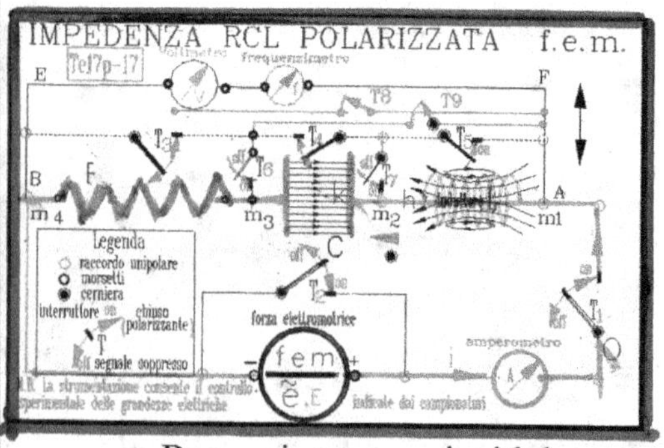

Per un ingresso cisoidale a componenti R-C-L in regime stazionario (nello stato zero) si definisce un integrale particolare a mezzo l'integrale definito: $f(s)=\int_0^{+\infty}e^{-st}f(t)dt$ (6) .Nella **Fig 17** è rappresentata una rete anulare di parametri R-CL. Posto ora l'ingresso in forma $u(t)=U_o\,e^{-\sigma t}\sin(\omega t+\alpha)$, che si particolarizza

consider the corripondenza:

$Y_o(t)=W(s)U_o(t)$ (1)

$Y_p(t)=Y_oe^{\sigma t}\sin(\omega t+\alpha+\varphi)$ (2)

The (2) Yp (t) is a **integral particular** if we consider as a function input

$u(t)=U_oe^{\sigma t}\sin(\omega t+\alpha)$ (3)

You can get the (2) with the symbolic method a particular integral of the equation in place of dif-ferenzuiale **pg 45** [I] and [II]. When the input is of the type (2) of generalized frequency $s=\sigma+j\omega$ (a) in phase with the argument φ of W (s). The calculation is denoted by $Z(s)=\omega L$ and $Y(s)=\omega C$ respectively the impedance and admittance symbolic , **Fig3**

The knowledge F.dT is sufficient pe r determine the $\phi(t)$, the phase shift. that emerges when one sets: $\dot W(s)=W(s)\,e^{j\phi t}$ (4), being W (s) the form of W (s).

The particular integral (2) in notation simbplic presents the magnitudes of input and output in **fig17** the e.m.f. $\tilde e(t)=Em\sin(\omega t+\alpha)$ is the input u (t) in real time $\dot U(s)=Uo\,e^{st}$ (5)

[a] The integral of PSLapalace (1749-1827)

For an input cisoidale R-C-L components in steady-state (in the zero state) is defined as a particular integral by the integral definite:

$$f(s)=\int_0^{+\infty}e^{-st}f(t)dt \qquad (6)$$

In **Fig 17** is shown a network annular parameters R-CL. Now place the input in the form $u(t)=Uo\,e^{-\sigma t}\sin(\omega t+\alpha)$, which is particularized

La Fig11 rappresenta il primo insieme di tagli della R.E. precedente. Il taglio della quale è attivato ponendo l'interruttore T_7 in off line. La soluzione in corrente con il metodo simbolico è possibile scrivenm due equazioni di maglia. Si supponga che la impedenza sL_6 e le ammettenze sC_2,sC_3,sC_4 nello stato zero[(che sono funzioni di trasferimeto ingresso u(t)e uscita y(t) , in regime cisoidale $s=\sigma+j\omega$ (a)] per $t<to=0$. Se ora , istante iniziale, to=0. si chiudono T_E e T_1 le maglie [MI],[MII].[MIII] ,

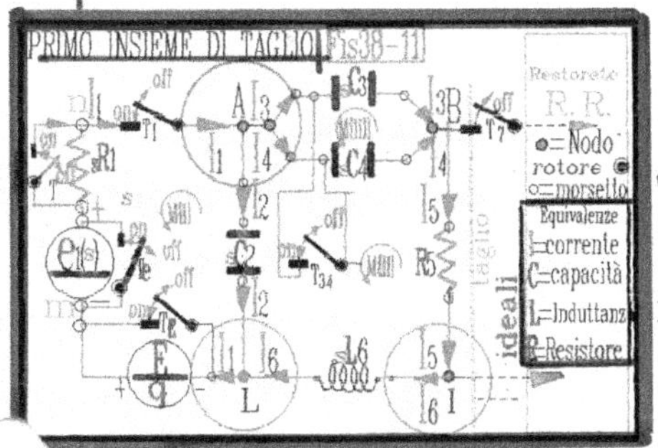

per $t \geq 0$ risultino polarizzate dall'ingresso sinusoidale

$$e_1(t)=E_m \sin(\omega t+\alpha) \quad (1) \leftrightarrow e_1(s)=\frac{E_m}{R_5} e^{s.t} \quad (2)$$

Essendo **Em** la ampiezza del segnale di ingresso **U**(t) e ωla pulsazione a regime ed α la fase della f.e.m .(1)

[a]Le equazioni di maglia e dei nodi della R.E. Fig11

Indichiamo con $\dot{U}=U_o e^{st}$ (2) [Ricordando la funzione di trasferimento((5)pg-58)]possiamo scrivere le tre equazioni algebriche di maglia:

Maglia [I] : $U_o e^{st}+sC_2=0$ (2° **Kirchoff** (3)

Maglia [II] : $sC_2-sL_2-R_5-sC_4=0$ " (4)

Maglia [III] : $sC_4-sC_3=0$ " (5)

Equazioni di nodo :A→ $I_1(s)-I_3(s)-I_4(s)-I_2(s)=0$ " (6)

positive se entranti	: B→ $I_3(s)+I_4(s)-I_5(s)=0$ " (7)
negative se uscenti	: I → $I_5(s)-I_6(s)=0$ " (8)
Per valori discordi.	: L→ $I_6(s)+I_2(s)-I_1(s)=0$ " (9)

Ricordiamo con le corrispondenze simboliche a colori non hanno significato fisico, salvo se si dichiara. Inoltre i segni ± algebrici sono positivi se la maglia, percorsa in senso antiorario e le correnti nei nodi sono entranti nei nodi, negative in senso opposto. Questo perchè è impossibile prevedere quale sia il verso da attribuire alle correnti in uscita. Allora se queste sono algebrico positivo il verso è quello supposto in caso contrario va cambiato. In breve il segno effettivo deve essere tale da soddisfare identicamente le 9 equazioni delle correnti. Innoltre per le tensioni il segno attribuito

The Fig11 represents the first set of cuts of the RE previous year. The cutting of such is activated by placing the switch in T_7 off line.

The solution in the current with the symbolic method is possible scrivenm two equations of knitting. Assume that the impedance sL_6 and admittances sC_2,sC_3,sC_4 in the zero state [(which are functions of trasferimeto input u (t) andoutput y (t), in regime cisoidal $s=\sigma+j\omega$ (a)] for $t<to=0$. If time, start time, to = 0. close T_E and T_1 meshes [MI], [MII]. [MIII]

, For t = 0 prove polarized entrance sinusoidal

$$e_1(t)=E_m \sin(\omega t+\alpha) \quad (1) \leftrightarrow e_1(s)=\frac{E_m}{R_5} e^{s.t} \quad (2)$$

Being **Em** the amplitude of the input signal **U** (t) and ωla pulsation regime and the phase of the fem. (1)

[a]**The equations of the nodes of the mesh and RE**Fig11 Denote by $\dot{U}=U_o e^{st}$ (2) [Recalling the transfer function ((5) pg-58)] we can write the three algebraic equations of mesh:

Maglia [I] : $U_o e^{st}+sC_2=0$ (2° **Kirchoff**) (3)

Maglia [II] : $sC_2-sL_2-R_5-sC_4=0$ " (4)

Maglia [III] : $sC_4-sC_3=0$ " (5)

Equations of node :

Recall with matches symbolic color have no physical meaning,unless you declare it.

A→ $I_1(s)-I_3(s)-I_4(s)-I_2(s)=0$ (1° **Kirchoff**) " (6)

B → $I_3(s)+I_4(s)-I_5(s)=0$ " (7)

I → $I_5(s)-I_6(s)=0$ " (8)

L→ $I_6(s)+I_2(s)-I_1(s)=0$ " (9)

Moreover the algebraic signs are positive if the jersey, traveled counterclockwise and currents in the nodes are nodes that belong to the negative ones in the opposite direction. This is because it is impossible to predict what the to be allocated to the output currents. Then if these are positive al-gebraic the direction is that supposed to be chan - ged otherwise. such as to satisfy identically the 9

[a]Riprendiamo la R.E. di **Fig11** che come detto, rappresenta il primo insieme di taglio della **R.E.** Ci proponiamo di risolvere nel dominio del tempo le correnti dei lati . Il metodo consiste :

[1°]- Assumere la d.d.p. di un lato qualsiasi della R.E [2°]- Integrare ,rispetto alla corrente incognita del lato fra i limiti definiti{0⁻ , t} la d.d.p dell'ingresso u(t) [3°]Integrare per parti la uscita y(t) in corrente di lato . La rete è supposta in stato zero, ad esempio:

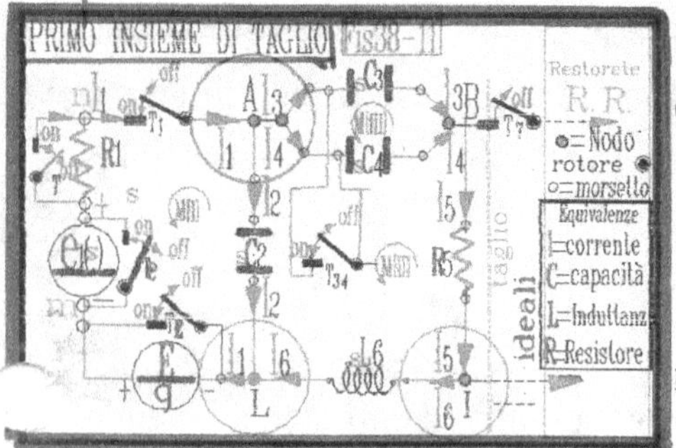

$V_{Ca}(0^-)=V_{Lb}(0^-)=0$ (1) A questo punto siamo in grado di iniziare le applicazioni per ogni singolo lato della **R.E** di **Fig11-Fi38** , di ogni maglia dopo posto off line l'interruttore T_7 Quindi si può iniziare

[b]Lato A↔B≡ Maglia [III] si trova in applicazione

$$V_{AB}(t)=\frac{1}{C_3}\int_0^t i_3(t)dt=\frac{1}{C_3}d\int_0^t i_3(t)dt=\frac{1}{C_3}[i_3(t)]_{-0}^{t}=\frac{1}{C_3}[i_3(t)-i_3(0-)]=\frac{i_3(t)}{C_3} \quad (2)$$

$$V_{BA}(t)=\frac{1}{C_4}\int_0^t i_4(t)dt=\frac{1}{C_4}d\int_0^t i_4(t)dt=\frac{1}{C_4}[i_4(t)]_{-0}^{t}=\frac{1}{C_4}[i_4(t)-i_4(0-)]=\frac{i_4(t)}{C_4} \quad (3)$$

[c] Lato A↔L≡ Maglia [I]

$$V_{AL}(t)=\frac{1}{C_2}\int_0^t i_2(t)dt=\frac{1}{C_2}d\int_0^t i_2(t)dt=\frac{1}{C_2}[i_2(t)]_{-0}^{t}=\frac{1}{C_2}[i_2(t)-i_2(0-)]=\frac{i_2(t)}{C_2} \quad (4)$$

$$V_{LA}(t)=R_1\int_0^t i_1(t)dt=R_1 d\int_0^t i_1(t)dt=R_1[i_1(t)]_{-0}^{t}=R_1[i_1(t)-i_1(0-)]=R_1 i_1(t) \quad (5)$$

[d] Lato B↔I≡ Maglia [I]

$$V_{BL}(t)=R_5\int_0^t i_5(t)dt=R_5 d\int_0^t i_5(t)dt=R_5[i_5(t)]_{-0}^{t}=R_5[i_5(t)-i_5(0-)]=R_5 i_5(t) \quad (6)$$

[e] Lato I↔L≡ Maglia [III]

$$V_{IL}(t)=L_6\int_0^t i_6(t)dt=L_6 d\int_0^t i_6(t)dt=L_6[i_6(t)]_{-0}^{t}=L_6[i_6(t)-i_6(0-)]=L_6 i_6(t) \quad (7)$$

Abbiamo risolto il problema del calcolo delle correnti legate agli ingressi di ogni singolo lra ingresso V(t) in termini di differenza di potenziale della uscita in corrente i(t) del circuito di **Fig 11**

[a] Resume the R.E. Fig11

of which as said, represents the first set of cutting R.E. We propose to solve the time-domain branch currents. The method consists of:

[1°] - Take the ddp of any side of the RE

[2°]- Integrate with respect to the current uncertainty of the side between the limits defined {0 -, t} the d.d.p. input u(t)

[3°] Integrating by parts the output y (t) current side. The network is supposed in the zero state, for example: $V_{Ca}(0-) = V_{Lb}(0-) = 0$ (1)

At this point we can begin the applications for each side of the **R.E.** of **Fig11-Fi38**, of each jersey after place off line switch T_7. With the Rest of this network is disconnected and then you can then complete the process for how many cuts you want.

[b] Side A↔B≡ Jersey [III] is located under:

$$V_{AB}(t)=\frac{1}{C_3}\int_0^t i_3(t)dt=\frac{1}{C_3}d\int_0^t i_3(t)dt=\frac{1}{C_3}[i_3(t)]_{-0}^{t}=\frac{1}{C_3}[i_3(t)-i_3(0-)]=\frac{i_3(t)}{C_3} \quad (2)$$

$$V_{BA}(t)=\frac{1}{C_4}\int_0^t i_4(t)dt=\frac{1}{C_4}d\int_0^t i_4(t)dt=\frac{1}{C_4}[i_4(t)]_{-0}^{t}=\frac{1}{C_4}[i_4(t)-i_4(0-)]=\frac{i_4(t)}{C_4} \quad (3)$$

[c] Side A↔L≡ Jersey [I]

$$V_{AL}(t)=\frac{1}{C_2}\int_0^t i_2(t)dt=\frac{1}{C_2}d\int_0^t i_2(t)dt=\frac{1}{C_2}[i_2(t)]_{-0}^{t}=\frac{1}{C_2}[i_2(t)-i_2(0-)]=\frac{i_2(t)}{C_2} \quad (4)$$

$$LA(t)=R_1\int_0^t i_1(t)dt=R_1 d\int_0^t i_1(t)dt=R_1[i_1(t)]_{-0}^{t}=R_1[i_1(t)-i_1(0-)]=R_1 i_1(t) \quad (5)$$

[d] Side L↔A≡ Jersey [I]

$$V_{BL}(t)=R_5\int_0^t i_5(t)dt=R_5 d\int_0^t i_5(t)dt=R_5[i_5(t)]_{-0}^{t}==R_5[i_5(t)-i_5(0-)]=R_5 i_5(t) \quad (6)$$

[e] Side I↔L≡ Jersey [III]

$$V_{IL}(t)=L_6\int_0^t i_6(t)dt=L_6 d\int_0^t i_6(t)dt=L_6[i_6(t)]_{-0}^{t}=L_6[i_6(t)-i_6(0-)]=L_6 i_6(t) \quad (7)$$

ave solved the problem the determination of the currents related to the inputs of each side between input V(t)in terms of the potential difference and the output current i(t) of the circuit of **Fig 11**

Le soluzioni in corrente calcolate in precedenza possono essere espresse direttamente mediante le tensioni ai terminali dei lati

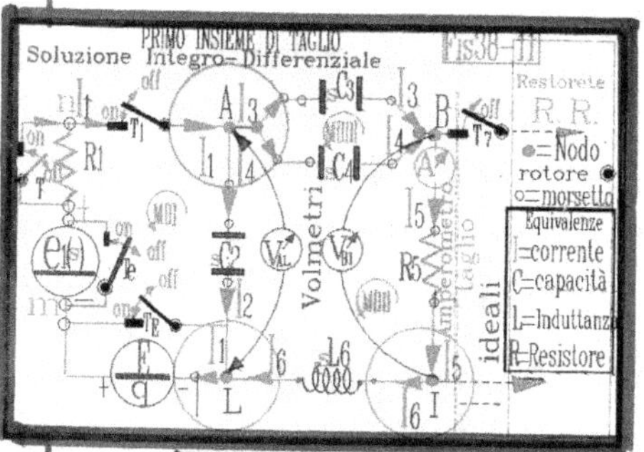

$$i_1(t)=\frac{E_m}{R_1}\sin(\omega t+\alpha) \quad (1)$$

$$i_2(t)=\frac{V_L(t)-V_A(t)}{C_2} \quad (2)$$

$$I_3(t)=\frac{V_B(t)-V_A(t)}{C_3} \quad (3)$$

$$I_4(t)=\frac{V_B(t)-V_A(t)}{C_4} \quad (4)$$

La verifica della d.d.p. ai

$$I_5(t)=\frac{V_I(t)-V_B(t)}{R_5} \quad (5)$$

morsetti di ogni lato viene

$$I_6(t)=L_6[V_L(t)-V_I(t)] \quad (6)$$

fatta con un voltmetro

Nella Fig11 si è indicato il colegamento off line di due voltmetri Precisamente V_{AL} posto in derivazione ai morsetti A ed L e V_{BI} derivato rispetto ai morsetti B ed I, con inserito in linea un amperometro.

[a] A numerical example

Dati del problema

N.B. Sono i valori dei componenti **R-C-L** presenti nella rete e sono indicati in unità Ohm (Ω) Allora $R_1(40\Omega)$, $R_5(50\Omega)$, $C_2(-20\Omega)$, $C_3(-40\Omega)$, $C_4(-50\Omega)$ $L_6(40\ \Omega)$ e1(t)

[b] Domande e risposte in corrente

[1]Quale corrente $i_1(t)$ attraversa il genetatore di tensione? Dalla $i_1(t)=\frac{E_m}{R_1}\sin(\omega t+\alpha)$ (1) sapendo il valore di $R_1=40$ e supposto che $e_1(t)$ sia definita da : $e(t) = E_m \sin(\omega t + \alpha) = 100 \sin(\omega t + \alpha)$ possiamo scrivere $i_1(t) = (100:40) \sin(\omega t + \alpha) = 2{,}5 \sin(\omega t+\alpha)$

La corrente che attraversa il generatore ha ampiezza di **2,5 Ampere** .Se abbiamo chiuso all'istante to=0 l'interruttore T_1 la fase $\alpha=0$ Quindi $i_1(t) = 2{,}5 \sin(\omega t)$

[2] Quale valore assume la corrente $i_2(t)$? per $t \geq 0$ Per $i_2(t)=\frac{V_L(t)-V_A(t)}{C_2}$ se si legge nel Voltmetro V_{LA}=150 Volt ? sapendo che C_2=-20Ω Si trova $i_2(t)= (150:-20)\sin(\omega t) = -7{,}5 \sin(\omega t)$ Se si legge al voltmetro V_{IB}=150 Volt si ha $i_5(t)= 3 \sin(\omega t)$. Con **3** la ampiezza I in Fig11

solutions in current calculated previously can be espresses directly by the voltages at the terminals of the sides

$$i_1(t)=\frac{E_m}{R_1}\sin(\omega t+\alpha) \quad (1)$$

$$i_2(t)=\frac{V_L(t)-V_A(t)}{C_2} \quad (2)$$

$$I_3(t)=\frac{V_B(t)-V_A(t)}{C_3} \quad (3)$$

$$I_4(t)=\frac{V_B(t)-V_A(t)}{C_4} \quad (4)$$

The verification of d.d.p. to

$$I_5(t)=\frac{V_I(t)-V_B(t)}{R_5} \quad (5)$$

terminal of each side is

$$I_6(t)=L_6[V_L(t)-V_I(t)] \quad (6)$$

made with a voltmeter In Fig11 is indicated the colegamento off line of two voltmeters Precisely V_{AL} place in shunt to the terminals A and L and V_{BI} derivative with respect to the terminals B and I, among which in on line is inserted an ammeter.

[a] A numerical example

Data of the problem

N.B. **R-C-L** are the values of the components present in the network and are indicated in units Ohm () Then R1 (40), R5 (50W), C2 (-20W), C3 (-40W), C4 (-50W) L6 (40 W) e1 (t)

[b] Questions and answers on current

[1] What current i_1 (t) crosses the genetatore of tension? From $i_1(t)=\frac{E_m}{R_1}\sin(\omega t+\alpha)$ (1) knowing the value of $R_1 = 40$ and assumed that e1 (t) is defined by: $e(t) = E_m \sin(\omega t + \alpha) = 100 \sin(\omega t + \alpha)$ we can write i1 (t) = (100:40) sin (wt a) = 2.5 sin (wt a) The current through the generator has an amplitude of **2.5 Amps**. If we closed instantly to = 0 wich interupter T_1 is phase $\alpha = 0$ Then $i_1(t) = 2.5 \sin(\omega t)$ [2] Which value assumes the current i_2 (t)? from $t \geq 0$ .For $i_2(t)=\frac{V_L(t)-V_A(t)}{C_2}$ if you read the voltmeter V_{LA} = 150 Volt? knowing that C_2 =-20Ω is located i2 (t) = (150: -20). sin (wt) = =-7.5 sin (wt). If the voltmeter reads V_{IB} = 150 volts i5 (t) = 3 sin (wt). Wich I amplitude in Fig11

Definizione. Un bipolo è inerte se sotto la d.d.p. ΔV **per**$\{-\infty\leftarrow t\rightarrow+\infty\}$ **è i(t)=0**.Oppure a morsetti * aperti per $\Delta V(t)=0$. In tali condizioni non può essere attraversato da corrente se ai suoi morsetti non si applica ΔV.

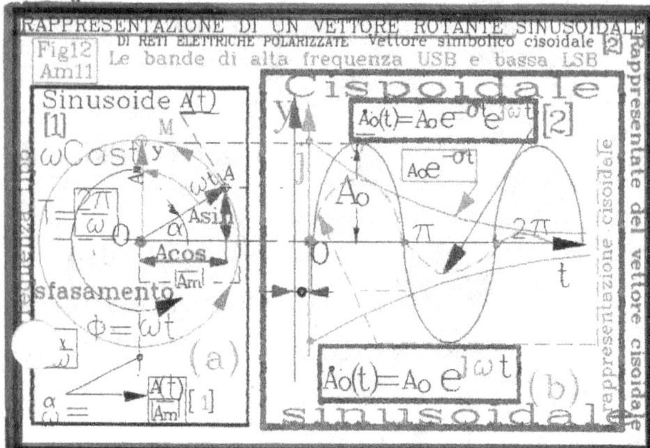

Soddisfatte queste condizioni d'ingresso la evoluzione è, se $\Delta V\neq 0$:

[a]- Formato Tempo variante rappresentato da una sinusoide (b) o da una cisoide [2], grandezze fisiche simboliche usate nelle reti elettriche con ingressi in tensione ed uscite in corrente.

In regime smorzato , $A(t)= A_o e^{-j\sigma(t)}$ (1) curva tratteggiata corrisponde alla presenta di uno o più condensatori caricati $qC=V$ in $t<t_o=0$ (stand by)

Se doppio doppio indutore in amplificazione all'uscita Vu(t) , del resto anche i componenti RCL

$Vu(t)\leftrightarrow Vi(t)$ **uscita -ingresso:**

[b]-Polinomi derivazionali

$$b_n\frac{d^n v}{dt^n}+b_{n-1}\frac{d^{n-1} v}{dt^{n-1}}+....+b_h\frac{d^h v}{dt^h}+....+b_1\frac{dv}{dt}+b_o=$$
$$=a_m\frac{d^m i}{dt^m}\ a_{m-1}\frac{d^{m-1} i}{dt^{m-1}}+....+a_h\frac{d^h i}{dt^h}+....+a_1\frac{di}{dt}+a_{oi} \qquad (1)$$

. Considerati sia in regime sinusoidale che cisoidale per bipoli **inerti**(RCLM) e **normali**,essendo le costanti $b\leftrightarrow a$ associate alle derivate Nella Fig7 abbiamo rappresentato un circuito anulare ,dotato di bipoli ideali e perfetti quali **C** ed **L** ai capi dei morsetti , Se si chiude l'interruttore T_1 all'istante $t=t_o=0$ possiamo leggere la evoluzione delle tensioni dovute alle cadute di potenziale, passando dallo stato zero allo stato attivo della corrente. Se poi si chiude Tc e si apre T1 può restare intrappolata una carica **q** e **C** quindi in stand by è un generatore a tempo determinato da **q**

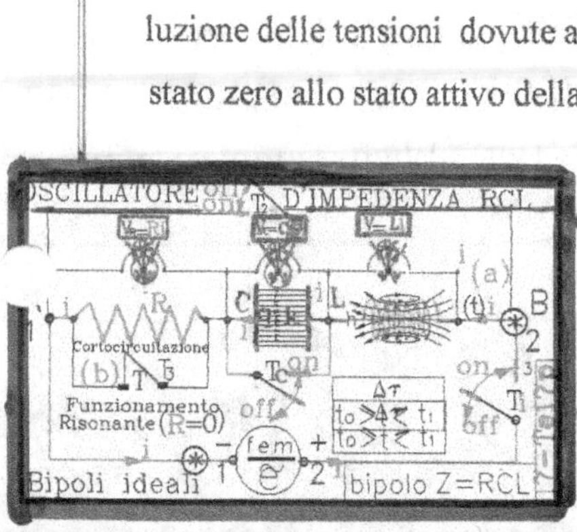

Definition. If an element is inert under the **d.d.p** $\Delta V\}$is i(t) = 0 in to terminals for $\{-\infty\leftarrow t\rightarrow+\infty\}$ is open for $\Delta V(t) = 0$.

In such conditions may not be current if atttraversato from its terminals does not apply ΔV. Meet all of these conditions of entry and the evolution is, if DV 0:

[a] - Time Format variant

represented by a sine wave (b) or a cisoide **[2]** the physical quantities used in the symbolic power grids with voltage inputs and current outputs.

In regime damped, $A(t)= A_o e^{-j\sigma(t)}$ (1) corresponds to the dashed curve presents one or more capacitors loaded in $qC = V$ in $t < t_o = 0$ (stand by) If double inductpr in amplification exit Vu (t), so are the components RCL

$Vu (t) \leftrightarrow Vi (t)$ **exit-entry:**

[b]-polynomials derivational.

$$b_n\frac{d^n v}{dt^n}+b_{n-1}\frac{d^{n-1} v}{dt^{n-1}}+....+b_h\frac{d^h v}{dt^h}+....+b_1\frac{dv}{dt}+b_o$$
$$=a_m\frac{d^m i}{dt^m}+a_{m-1}\frac{d^{m-1} i}{dt^{m-1}}+...+a_h\frac{d^h i}{dt^h}+...+a_1\frac{di}{dt}+a_{oi}\ (1$$

Considered both in the sinusoidal dipoles that cisoidale for **aggregates** (RCLM) and **normal,** being associated with the derived constants ba In Fig7 we have represented an annular circuit with dipoles ideal and perfect such as **C** and **L** at the terminal ends, If you close the 'switch **T** at $t = t_o = 0$ we can interpret the evolution of the stresses due by potenale of passing from the zero state to the active state of the current. If then the switch is closed and opens Tc T_1 can be trapped in a charge **q** and **C** then stand by generator is a time determined by the charge **q**

[a]-Il significato dei Polinomi derivazionali

$$b_n \frac{d^n v}{dt^n} + b_{n-1}\frac{d^{n-1} v}{dt^{n-1}} + \dots + b_h \frac{d^h v}{dt^h} + \dots + b_1 \frac{dv}{dt} + b_o \equiv$$

(ingresso) → (uscita)

$$\equiv a_m \frac{d^m i}{dt^m} + a_{m-1}\frac{d^{m-1} i}{dt^{m-1}} + \dots + a_h \frac{d^h i}{dt^h} + \dots + a_1 \frac{di}{dt} + a_o \quad (1)$$

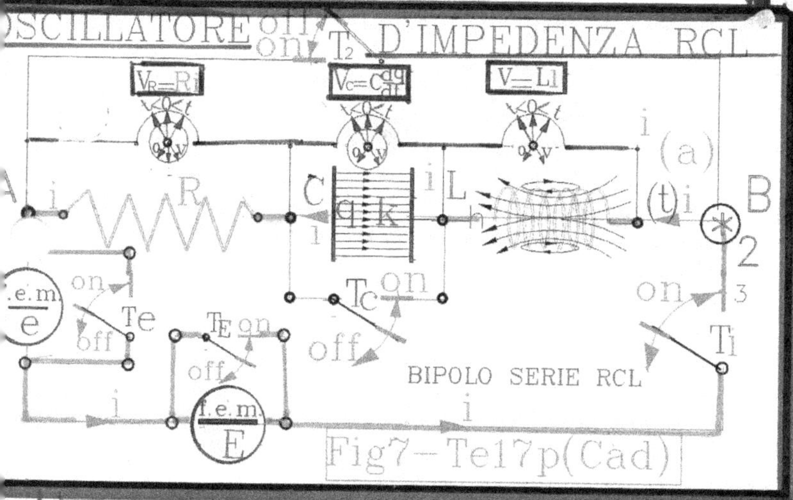

Fig7–Te17p(Cad)

Si supponga la rete di **Fig.7** nello stato zero e di chiudere, all'istante $t=t_o=0$ T_1 e T_E. Questo significa che la impedenza $\dot{Z} = R + j(L\omega - \frac{1}{\omega C})$ (2) è polarizzata dalla **f.e.m**. on line. Al solito definita dalla equazione: $e(t)=E_m \sin(\omega t + \alpha)$ (3), che rappresenta l'ingresso in tensione nel circuito R-C-L La presenza della pulsazione $\omega=2\pi f$ (4) relativa al secondo membro della (1) La legge della polarizzazione assicura che per ogni componente di ogni singolo lato è indotta una frequenza **f** relativa alla (3). Funzione ciclica derivabile di ordine 1,2....n:

$$b_n \frac{d^n v}{dt^n} + b_{n-1}\frac{d^{n-1} v}{dt^{n-1}} + \dots + b_h \frac{d^h v}{dt^h} + \dots + b_1 \frac{dv}{dt} + b_o \equiv$$

$$\equiv b_o - \omega\cos(\omega t + \alpha) + \omega^2 \sin(\omega t + \alpha) - \omega^3 \cos(\omega t + \alpha) + \dots + \omega^n \sin(\omega t + \alpha) \quad (2)$$

Dal fatto che la frequenza **(4)** è una costante della pulsazione della f.e.m. di ingresso è dimostrato che i coefficienti delle derivate della tensione ai capi dei bipoli, sono le ω costante dell'ingresso della (3) di tutti i bipoli della **R.E**.

[a]-The meaning of polynomials derivational

$$b_n \frac{d^n v}{dt^n} + b_{n-1}\frac{d^{n-1} v}{dt^{n-1}} + \dots + b_h \frac{d^h v}{dt^h} + \dots + b_1 \frac{dv}{dt} + b_o \equiv$$

(aut) → (put)

$$\equiv a_m \frac{d^m i}{dt^m} + a_{m-1}\frac{d^{m-1} i}{dt^{m-1}} + \dots + a_h \frac{d^h i}{dt^h} + \dots + a_1 \frac{di}{dt} + a_o i \quad (1)$$

Suppose the network of **Fig.7** in the zero state and clos-ing, at $t = t_o = 0$ **T1** and T_E.

This means that

the impedance $\dot{Z} = R + j(L\omega - \frac{1}{\omega C})$ (2) is polarized by the **f.e.m** on line. Al usually defined by the equation: $e(t) = E_m \sin(\omega t + \alpha)$ (3)

that represents the input voltage in the circuit RCL

The presence of the pulsating-tion $\omega = 2\pi f$ (4) relative to the second member of (1)

the law ensures that the polarization of each component for today sincolo side is induced at a frequency f relative to (3).

Since this is a loop it admits derivatives up to order n

$$b_n \frac{d^n v}{dt^n} + b_{n-1}\frac{d^{n-1} v}{dt^{n-1}} + \dots + b_h \frac{d^h v}{dt^h} + \dots + b_1 \frac{dv}{dt} + b_o \equiv$$

$$\equiv b_o - \omega\cos(\omega t + \alpha) + \omega^2 \sin(\omega t + \alpha) - \omega^3 \cos(\omega t + \alpha) + \dots + \omega^n \sin(\omega t + \alpha) \quad (2)$$

From the fact that the frequency (4) is a constant pulsation of the emf input is shown that the coefficients of the derivative of the voltage across the bipoles, w are the constant input of the (3) of all the circuits of **R.E.**

KIRCHHOFF 1824-1887

J.C. MAXWELL 1831-1892

H. HERTZ 1857-1894

[a]-Le trasformate di Laplace(LT)

Reintroduciamo(pg-58) La formula atta a trassformare una f(t)nella corripondente f(s), con s=jωt è:

$$LT[f(t)]=f(s)=\int_0^{\infty} f(t)e^{-st}dt \quad (1)$$

Naturalmente seLT[f(t)] =f(s) .Non tutte le f(t) sono **Laplace** trasformabili Esempio di una f(t) trasformabile secondo la(1) Sia f(t)=1 .Si calcoli la sua LT[f(t)=1]= f(s)

[b]Calcolo della LT della funzione f(t)=1

Per la (1) $\int_0^{\infty} f(t)e^{-st}dt \rightarrow \int_0^{\infty} e^{-st}dt$ (1_1) Ora derivando e^{-st} si trova: la soluzione secondo il procedimento a seguire.

$\frac{d}{dt}e^{-st}=-se^{-st}$ (1_2) da cui $d\,e^{-st} = -s\,e^{-st}\,dt$ (1_3)

Anche $-\frac{1}{s}d\,e^{-st} = e^{-st}dt$ (1_4) Introdotta la (1_4) nella(1_1)si ha : $\int_0^{\infty} e^{-st}dt = \int_0^{\infty}\frac{de^{-st}}{-s} = \frac{1}{s}$ (2) Infatti anti-trasformando la $f(s)=\frac{1}{s}$ della f(t)=1 si ottiene:

$$LT^{-}[\tfrac{1}{s}] = \int_0^{\infty}\frac{d}{dt}e^{-st}/-s = \int_0^{\infty} e^{-st} = [e^{-st}]_0^{\infty} = 1 \quad (3)$$

[c] La opportunità delle tabulazioni

Se per calcolare la LT (2) della semplice f(t)=1 e dato in largo uso applicativo nel calcolo simbolico delle reti elettriche non c'è testo della materia che non riporti le trasformate LT[f(t)]↔:f(s) (3) e le antitrasformate LT^{-}[f(s)]= f(t) (4) tabulate. Allo scopo di supportare il calcolo simbolico oggetto del seguente capitolo riportiamo alcune semplici LT[f(t)]

TRASFORMATE DI LAPLACE ELEMENTARI

f(t)	Tab-a/Am13		LT{f(t)} = f(s)	
1	1		$\frac{1}{s}$	s>0
2	t		$\frac{1}{s^2}$	s>0
3	t^n		$\frac{n!}{s^{n+1}}$	s>0
4	e^{at}		$\frac{1}{s-a}$	s>a
5	sin(ta)		$\frac{a}{s^2+a^2}$	s>a
6	cos(ta)		$\frac{s}{s^2+a^2}$	s>0
7	sinh(ta)		$\frac{a}{s^2-a^2}$	s >|a|
8	cosh(ta)		$\frac{s}{s^2-a^2}$	s >|a|

[a]-The Laplace transforms (LT)

For (pg-58) The formula designed to trassfor a sea-(t) in the penfriend f (s), with s = jωt is:

$$LT[f(t)]=f(s)=\int_0^{\infty} f(t)e^{-st}dt \quad (1)$$

Of course if LT [f (t)] =f(s)
Not all f (t) are Laplace transformable Example of f (converts the for (1) Not all f (t)
Let f (t) = 1. Compute its LT [f (t) = 1] = f (s)

[b] Calculation of the LT of the function f (t) = 1

For the (1) $\int_0^{\infty} f(t)e^{-st}dt \rightarrow \int_0^{\infty} e^{-st}dt$ (1_1)
Now resulting e^{-st} is: the soluzion wich the course a follow.

$\frac{d}{dt}e^{-st}=-se^{-st}$ (1_2) da cui $d\,e^{-st} = -s\,e^{-st}\,dt$ (1_3)

To : $-\frac{1}{s}d\,e^{-st} = e^{-st}dt$ (1_4) well-knpwn (1_4) i (1_1). Is : $\int_0^{\infty} e^{-st}dt = \int_0^{\infty}\frac{de^{-st}}{-s} = \frac{1}{s}$ (2)

In fact, antitransforming f (s) = the f (t) = 1 is obtained:

$$LT^{-}[\tfrac{1}{s}] = \int_0^{\infty}\frac{d}{dt}e^{-st}/-s = \int_0^{\infty} e^{-st} = [e^{-st}]_0^{\infty} = 1 \quad (3)$$

[c] The opportunity of tabs

If you calculate the LT (2) thr simple f (t) = 1 and given in wide use in the application of symbolic computation grids there is no text of matter that does not show that the transformed LT[f(t)]↔:f(s) (3) and the inverses LT^{-}[f (s)] = f (t) (4) simple tabulated.

In order to support symbolic computation object of the following section offers some LT [f (t)]

P.S. LAPLACE 1749-1827

APPLICAZIONI DI L.T↔ R.E

[a]Uso delle L.T per la soluzione delle R.E.

Questo metodo si aggiunge a quello delle equazioni differenziali di **pg-48** ed a quello integro--differenziale di **pg-75** più adatto del precedente quando la **R.E.** ha molte maglie.In questo caso è particolarmente adatto il metodo simbolico con gli operatori di Laplace,per lo studio delle reti in regime quasi stazionario . E' opportuno tenere presente che il calcolo operatoriale :[1] Perchè possa **LT**[f(t)] =**f(s)** (1) occorre, ma non è sufficiente che f(t) sia nulla per valori di t <0. La condizione è sufficiente se la rete si trova nello stato zero

[2]-Per la trasformazione inversa **LT⁻[f(s)]**=f(t) deve essere f(t) tale che sia nulla per t<0

[3] Se le precedenti [1] e [2] sono verificate per lo stato della rete allora le tensioni e le correnti sono tutte nulle per t<0. Le reti vanno pensate tutte nulle identicamente per t<0.

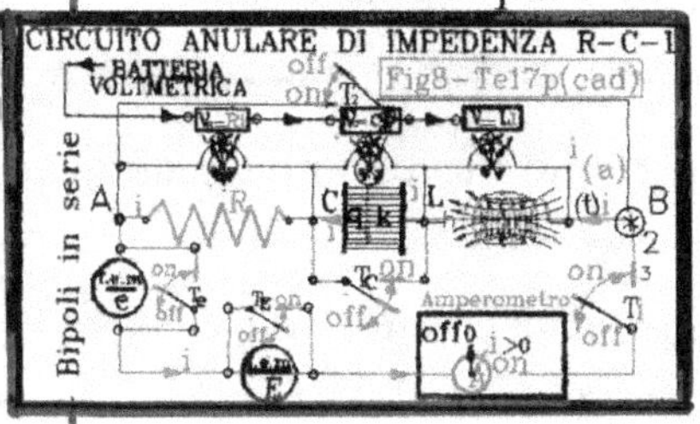

Nella Fig8 abbiamo riportato la batteria di voltmetri in derivazione a componenti considerati in alternativa aperti o chiusi con ingressi applicati nella forma f(t) δ-(t).La risposta y(t), per t ≥ 0, alla chiusura dei tasti T_1(T=abbreviazione di interruttore) sarà possibile anche quando per t<0 purchè la si riconduca a tensioni non nulle per t ≥ 0

[b] Dimostrazione della F.di T.

Nella **pg-71** è stata introdotta la funzione di trasferimeno relativa ad una certa uscita per un certo ingresso in forma $W(s)=\frac{Y(s)}{U(s)}$ (2) Si consideri la rete di Fig11-Fis38 già vista .La (2) relativa ad una certa uscita per un dato ingressoè uguale alla **LT** di ingresso e quella della uscita.Se si C.C. la **f.e.m. E** chiudendo T_E il resto rete di un primo taglio le tre Maglie sono alimentate dal solo generatore $e_1(s)$ Si supponga che R-C-L siano tempo invarianti,........

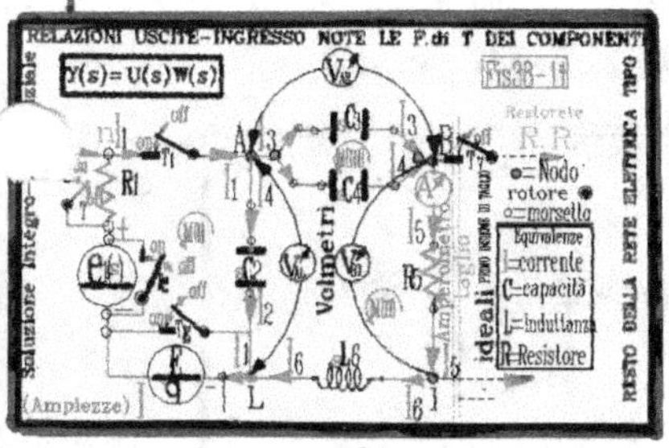

APPLICATIONS OF LT↔ RE

[a] Using LT for the solution of R.E

This method is added to that of the differential equations of **pg-48** and to the integro-differential-**pg-75** more appropriate than the previous year when the **R.E** has many maglie. In this case is particular lysuitable symbolic methods with the operators of Laplace, for the study of quasi steady-state networks.

E 'should keep in mind that the calculation operatorial:[1]

Why can **LT** [f (t)] = **f (s)** (1) is necessary, but it is not sufficient that f (t) is zero for values of t <0 . The condition is sufficient if the network is in the state zero

[2] For the reverse transformation **LT⁻[f(s)]** = f (t) must be f (t) such that it is zero for t <0

[3] If the above [1] and [2] are checked for the status of the network then the voltages and currents are all zero for t <0.

The networks are designed identically zero for all t <0. In dwg8 we have indications given battery voltmeters in derivation to com-ponents considered alternatively open or closed with inputs applied in the form f (t)-d (t). Response y (t), for t 0, the closure of keys T1 (T = short for switch) it will be possible even when for t <0 as long as the lead back to voltages other than zero for t <0

[b] Proof of F. Di T

In page-71 was introduced the function of transfer charger relative to a certain output for a certain input in the form $W(s)=\frac{Y(s)}{U(s)}$ (2) Consider the network of dwg11-Fis38 already seen.

La (2) relative to a certain output for a given ingressoè **LT** equal to the input and that of the C.C is out .Se the **f.e.m.** ,and closing T_E the rest of the network of a first cut of the three jerseys are powered by one generator e_1 (s)

Suppose that RCL are time-invariant ,. ....

Si consideri la rete anulare di **Fig17** a componenti R-C-L- lineari e tempo invarianti . Volendo polarizzare la impendenza totale $Z = R + j(L\omega - \frac{1}{C\omega})$ (1) con un solo **generatore**,ad esempio $e(t) = Em \sin(\omega t + \alpha)$ (2) si deve porre T_E off line Dopo ciò alla chiusura di T_i l'impendenza (1) è polarizzata. In fatti i Voltmetri evidenziano le d.d.p. ai terminali degli elementi lineari ed invarianti. In via sperimentale la uscita y(t) è rivelata dall'amperometro (on line) della corrente in uscita y(t) in relazione all'ingresso u(t) espresso dalla (2) . Ora se $u(t) \equiv 0$ per $t<to=0$ si può operare con il calcolo operatoriale .

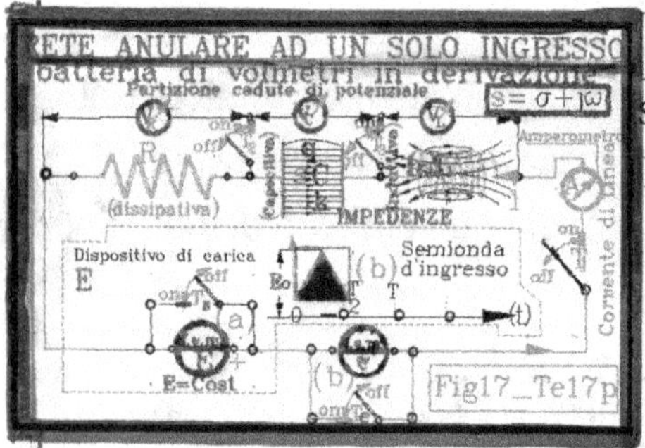

[a] Le applicazioni del calcolo operatoriale

In tempo reale la relazione Uscita-Ingresso sono definite, in termini derivazionali :

$$\sum_{i=0}^{n} a_i \frac{d^{(i)} y(t)}{d^i} = \sum_{i=0}^{m} b_i \frac{d^{(i)} u(t)}{dt^i} \quad (3)$$

.Infatti l'ingresso (2) è derivabile fino all'ordine voluto in quanto si può verificare la successione periodica in tempo reale:

$$\frac{d}{dt} Em \sin(\omega t + \alpha) = \omega Em \cos(\omega t + \alpha) \quad (4)$$

$$\frac{d^2}{dt^2} Em \sin(\omega t + \alpha) = -\omega^2 Em\sin(\omega t + \alpha) \quad (5)$$

$$\frac{d^3}{dt^3} Em \sin(\omega t + \alpha) = \omega^3 Em \cos(\omega t + \alpha) \quad (6)$$

Relativa al secono membro della (3) derivata fino all'ordine m e quindi con l'uscita y(t) relativa a ciascun componente della rete elettrica assegnata

[b] Le trasformate ed antitrasformate di Laplace

Ricordiamo che la f(s),pg-59(8), è la **LT** della funzione f(t). Se risulta $f(0^-)=0$ per $t<to=0$,la rete si trova nello stato zero .In pratica chiudendo tutti gli interruttori ,ad esempio della **Fig17** , l'amperomtero non rileva alcuna corrente in transito. come dire che la rete si trova nello stato inerte . Questo vale anche per reti multimaglie

G.B.FORIER

the network annular components **fig17** to R-C-L linear and time invariant. Wanting to polarize the total impedance $Z = R + j(L\omega - \frac{1}{C\omega})$ (1) with a single **generator**, for example, $e(t) = Em \sin(\omega t + \alpha)$ (2) it must be put off line T_E After that the closure of the impedance Ti (1) is polarized. In fact, the volmeter highlight the d.d.p. to the terminals of linear elements and invariants.

As an experiment the output y (t) is revealed by the ammeter , (online), output current y (t) in relation to the input u (t) expressed by (2). Now if $u(t) \equiv 0$ for $t < to = 0$ it can operate with the calculation operatorial.

[a]The application of the wich operator

Real-time input-output relationship are defined in terms derivational:

$$\sum_{i=0}^{n} a_i \frac{d^{(i)} y(t)}{d^i} = \sum_{i=0}^{m} b_i \frac{d^{(i)} u(t)}{dt^i} \quad (3)$$

. Fact the input (2) is differentiable up to the order-liked since you may experience periodic succession in real time:

$$\frac{d}{dt} Em \sin(\omega t + \alpha) = \omega Em \cos(\omega t + \alpha) \quad (4)$$

$$\frac{d^2}{dt^2} Em \sin(\omega t + \alpha) = -\omega^2 Emsin(\omega t + \alpha) \quad (5)$$

$$\frac{d^3}{dt^3} Em \sin(\omega t + \alpha) = \omega^3 Em \cos(\omega t + \alpha) \quad (6)$$

Relative to secono of (3) derived up to order me then with the output y (t) for each component of the electricity grid assigned

[b] The Laplace transform and inverse transforms

Recall that f (s), pg-59 (8), is the **LT** of the function f (t). If it is f (0 -) = 0 for t <to = 0, the network is in the zero state. Basically closing all switches, such as the **fig17**, the amperomtero does not detect a current in transit. as saying that the network is in the quiescent state. This also applies to networks multimaglie

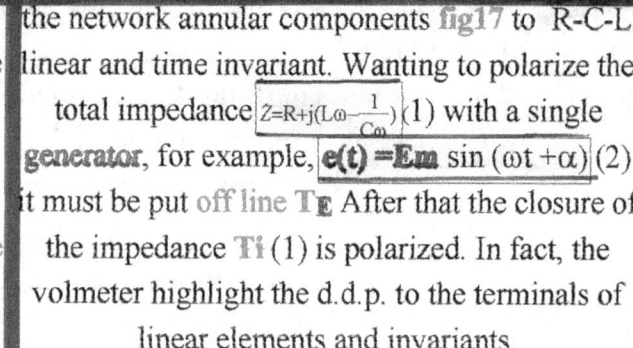

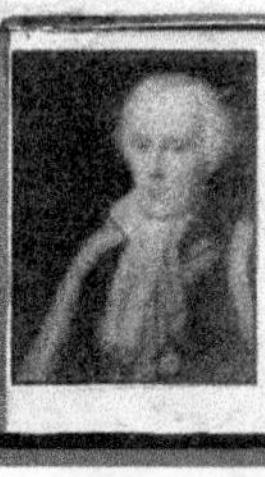

LAPLACE 1749–1827

www.ingramcontent.com/pod-product-compliance
Lightning Source LLC
Chambersburg PA
CBHW081048170526
45158CB00006B/1901
* 9 7 8 1 2 9 1 7 5 3 6 7 7 *